Autodesk Revit 2017 Architecture Certification Exam Study Guide

Elise Moss

Publications

SDC Publications
P.O. Box 1334
Mission, KS 66222
913-262-2664
www.SDCpublications.com
Publisher: Stephen Schroff

ISBN-13: 978-1-63057-085-9
ISBN-10: 1-63057-085-0

Printed and bound in the United States of America.

Preface

This book is geared towards users who have been using Revit for at least six months and are ready to pursue their official Autodesk certification. You can locate the closest testing center on Autodesk's website. Check with your local reseller about special Exam Days when certification exams can be taken at a discount.

I wrote this book because I taught a certification preparation class at SFSU. I also teach an introductory Revit class using my Revit Basics text. I heard many complaints from students who had taken my Revit Basics class when they had to switch to the Autodesk AOTC courseware for the exam preparation class. Students preferred the step by step easily accessible instruction I use in my texts.

This textbook includes exercises which simulate the knowledge users should have in order to pass the certification exam. I advise my students to do each exercise two or three times to ensure that they understand the user interface and can perform the task with ease.

I have endeavored to make this text as easy to understand and as error-free as possible…however, errors may be present. Please feel free to email me if you have any problems with any of the exercises or questions about Revit in general.

I have posted videos for some of the lessons from this text on my website at www.mossdesigns.com on the Tutorials page. The videos are free to view. Videos are also posted on YouTube – search for Moss Designs and you will find any videos for my book on that channel.

Exercise files can be accessed and downloaded from the publisher's website at: www.sdcpublications.com/downloads/978-1-63057-085-9

I have included two sample practice exams, one for the User Certification exam and one of the Professional certification exam. These may only be downloaded using a special access code printed on the inside over of the textbook. The practice exams are free and do not require any special software, but they are meant to be used on a computer to simulate the same environment as taking the test at an Autodesk testing center.

Please feel free to email me if you have any questions or problems with accessing files or regarding any of the exercises in this text.

Acknowledgements

A special thanks to Rick Rundell, Gary Hercules, Christie Landry, and Steve Burri, as well as numerous Autodesk employees who are tasked with supporting and promoting Revit.

Additional thanks to Gerry Ramsey, Will Harris, Scott Davis, James Cowan, James Balding, Rob Starz, and all the other Revit users out there who provided me with valuable insights into the way they use Revit.

Thanks to Zach Werner for the cover art work for this textbook along with his technical assistance. Thanks to Karla Werner for her help with editing and formatting to ensure this textbook looks proper for our readers.

My eternal gratitude to my life partner, Ari, my biggest cheerleader throughout our years together.

Elise Moss
Elise_moss@mossdesigns.com

Table of Contents

Introduction
FAQs on Getting Certified in Revit

The first day of class students are understandably nervous and they have a lot of questions about getting certification. Throughout the class, I am peppered with the same or similar questions.

Certification is done through a Certiport website. You can learn more at www.certiport.com/autodesk. The website currently has exams for Revit Architecture, Revit MEP Electrical, Revit MEP Mechanical and Revit Structure. You can be certified as a User and/or Professional. In order to be certified as a Professional, you have to pass both the User and the Professional exams. Once you are certified as a Professional, you only have to take the Professional exam to keep your certification up to date.

To pass the User exam, Certiport recommends 50 hours of practice prior to taking the exam. Both exams include multiple choice questions, fill in the blank, point and click, as well as questions that require use of the software (performance-based questions).

The Autodesk Certified User Exam is 30 questions and 50 minutes. This means you have an average of one and a half minutes per question. So, you need to be able to read and work quickly.

The Autodesk Certified Professional Exam is 35 questions and 120 minutes. This means you have an average of three and a half minutes per question. The performance based questions have more steps, so you still need to be able to work quickly.

I have indicated in the exercises which exercises are similar to questions which might appear in the User exam and which might appear in the Professional exam, so you can focus on those exercises which pertain to the exam you plan to take. Regardless, it is a good idea to spend some time building speed and figuring out which shortcuts you can use so you can complete tasks quickly and easily.

What are the exams like?

The first time you take the exam, you will create an account with a login. Be sure to write down your login name and password. You will need this regardless of whether you pass or fail.

If you fail and decide to retake the exam, you want to be able to log in to your account.

If you pass, you want to be able to log in to download your certificate and other data.

Autodesk certification tests use a "secure" browser, which means that you cannot cut and paste or copy from the browser. You should use the Alt-TAB shortcut key to flip from the secure browser to the software application to do each problem. If you go on Autodesk's site, there is a list of topics covered for each exam. Expect questions on the user interface, navigation, and zooming as well as how to place doors and windows. The topics include collaboration, creating Revit families, linking and monitoring files, as well as more complex wall families.

Exams are timed. This means you only have one to three minutes for each question. You have the ability to "mark" a question to go back if you are unsure. This is a good idea because a question that comes later on in the exam may give you a clue or an idea on how to answer a question you weren't sure about.

Exams pull from a question "bank" and no two exams are exactly alike. Two students sitting next to each other taking the same exam will have entirely different experiences and an entirely different set of questions. However, each exam covers specific topics. For example, you will get at least one question about BIM or Revit projects. You will probably not get the same question as your neighbor.

At the end of the exam, the browser will display a screen listing the question numbers, the answers you chose, and indicate any marked or incomplete questions. The questions themselves are not displayed on this page. Any questions where you forgot to select an answer will be marked incomplete with a capitol I. Questions that you marked will display the letter M. You can click on those answers and the browser will link you back directly to those questions so you can review them and modify your answers.

I advise my students to mark any questions where they are struggling and move forward, then use any remaining time to review those questions. A student could easily spend ten to fifteen minutes pondering a single question and lose valuable time on the exam.

Once you have completed your review, you will receive a prompt to END the exam. Some students find this confusing as they think they are quitting the exam and not receiving a score. Once you end the exam, you may not change any of your answers. There will be a brief pause and then you will see a screen where you will be notified whether you passed or failed. You will also see which questions you missed, listed by the question number. Again, you will not see any of the actual questions. However, you will see the topic, so you might see that you missed Question 4 and that question is on view properties.

You will have to rely on your memory to recall what question was about view properties. Some students might find this confusing because they didn't even know that particular question fit into the view properties category.

If you failed the test, you do want to note which categories you missed. For example, you might have missed a question on BIM, a question on view properties, and a question on constraints. Write these down. Then review those topics both in this guide and in the software to help prepare you to retake the exam.

It is helpful during the exam to have a piece of paper and a pencil so you can make notes.

You do not need a calculator for the exam and calculators are against the rules anyway.

How many times can I take the test?

You can take any test up to three times in a 12-month period. There is no waiting period between re-takes, so if you fail the exam on Tuesday, you can come back on Wednesday and try again. Of course, this depends on the testing center where you take the exam.

Do a lot of students have to re-take the test or do they pass on the first try?

About half of my students pass the exam on the first try, but it definitely depends on the student. Some people are better at tests than others. My youngest son excels at "multiple guess" style exams, which is what the user certification exam is. He can pretty much ace any multiple guess exam you give him regardless of the topic. Most students are not so fortunate. Some students find a timed test extremely stressful. For this reason, I have created a simulated version of the exam for my students simply so they can practice taking an online timed exam. This has the effect of "conditioning" their responses so they are less stressed taking the actual exam.

Some students find the multiple choice style extremely confusing, especially if they are non-native English speakers. There are exams available in many languages, so if you are not a native English speaker, check with the testing center about the availability of an exam in your native language.

Why take a certification exam?

The competition for jobs is steep and employers can afford to be picky. Being certified provides employers with a sense of security knowing that you passed a difficult exam that requires a basic skill set. It is important to note that the certification exam does not test your ability as a designer or drafter. The certification exam tests your knowledge of the Revit software. This is a fine distinction, but it is an important one.

If you pass the exam, you have the option of having your name published on Autodesk's website as a Certified Professional. Some employers and headhunters use that list to find a potential new hire. You do not have to have your name published if you don't want to be bothered.

How long is the certification good for?

Your certification is good for a particular year of software. It does not expire. If you pass and are certified for Revit 2017, you are always certified for Revit 2017. That said, you are not certified for Revit 2018 or Revit 2020. Most employers want you to be certified within a couple of years of the most current release, so if you wish to maintain your "competitive edge" in the employment pool, expect that you will have to take the certification exam every two to three years. I recommend students take an "update" class from an Autodesk Authorized Training Center before they take the exam to improve their chances of passing. Autodesk is constantly tweaking and changing the exam format. Check with your testing center about the current requirements for certification. The Certiport exams are no longer tied to a particular release year, so you can't specify which release you want to test on. You are always going to test on the current release.

How much does it cost?

The cost for the exam is posted on Autodesk's website. Autodesk sometimes charges students less than professionals, so if you are a student or unemployed, check to see if you qualify for a discount or a free exam. Members of AUGI, Autodesk User Group International, may also qualify for a discount or a free exam. Membership in AUGI is free. If you attend Autodesk University, some years the conference will allow attendees to take a certification exam for free or at a discount. It is worth it to check with the testing center to find out what criteria is necessary to qualify for a discount or free exam. Some testing centers offer "certification days" where users can take the exam at a discount on specific dates. Check with your local reseller or ATC to find out if there are any upcoming certification days available.

Can I take the exam at home?

No. Autodesk requires users to take the exam in a proctored setting. When I proctor an exam, I ask my students to show valid ID, such as a driver's license. This is to ensure that the certification process maintains a certain level of integrity and meaningfulness. Otherwise, a user could pay someone to take the exam for him.

Do I need to be able to use the software to pass the exam?

This sounds like a worse question than intended. Some of my students have taken the Revit classes, but have not actually gotten a job using Revit yet. They are in that Catch-22 situation where an employer requires experience or certification to hire them but they can't pass the exam because they aren't using the software every day. For those students, I advise some self-discipline where they schedule at least six hours a week for a month where they use the software – even if it is on a "dummy" project – before they take the exam. That will boost the odds in their favor.

What happens if I pass?

Because you are taking the test in a testing center, you want to be sure you wrote down your login information. That way when you get back to your home or office, you can login to Autodesk's certification center and download your certificate. You also can download a logo which shows you are a Certified User or Certified Professional that you can post on your website or print on your business card.

How many times can I log into Autodesk's testing center?

You can log in as often as you like. The tests you have taken will be listed as well as whether you passed or failed.

Can everybody see that I failed the test?

Autodesk is kind enough to keep it a secret if you failed the exam. Nobody knows unless you tell them. If you passed the test, people only see that information if you selected the option to post that result. Regardless, only you and Autodesk will know whether you passed or failed unless *you* choose to share that information.

Can I change my mind? I said I didn't want it posted if I passed and now I am looking for a job.

Yes, you are allowed to change your mind. Log on to the Autodesk testing site and email them using the Contact Link. Provide them with your user name, email address, and the test results you want posted. They will verify that you passed and update their database listing. This may take a week or longer. So, if you change your mind, let them know right away. Do not go on a job interview and say you are certified without bringing along your certificate as proof. Many employers will check your claim against Autodesk's database. A lot of employers now require a Certificate Number as proof that you are actually certified.

What if I need to go to the bathroom during the exam or take a break?

You are allowed to "pause" the test. This will stop the clock. You then alert the proctor that you need to leave the room for a break. When you return, you will need the proctor to approve you to re-enter the test and start the clock again.

How many questions can I miss?

Autodesk requires a passing score of 80%. Because the test is constantly changing, the number of questions can change and the amount of points each question is worth can also vary.

How much time do I have for the exam?

The User exam is 50 minutes and the Professional exam is 120 minutes.

Can I ask for more time?

You can make arrangements for more time if you are a non-English speaker or have problems with tests. Be sure to speak with the proctor about your concerns. Most proctors will provide more time if you truly need it. However, my experience has been that most students are able to complete the exam with time to spare. I have only had one or two students that felt they "ran out of time."

What happens if the computer crashes during the test?

Don't worry. Your answers will be saved and the clock will be stopped. Simply reboot your system. Get Revit launched again and let the proctor know when you are ready to start the exam again, so you can re-enter the testing area in your browser.

Can I have access to the practice exams you set up for your students?

I am including a practice exam for both the User and Professional versions in the exercise files with this text.

What sort of questions do you get in the exams?

Many students complain that the questions are all about Revit software and not about building design or the uniform building code. Keep in mind that this test is to determine your knowledge about Revit software. This is not an exam to see if you are a good architect or designer.

The exams have several question types:

One best answer – this is a multiple choice style question. You can usually arrive at the best answer by figuring out which answers do NOT apply.

Select all that apply – this can be a confusing question for some users because unless they know how *many* possible correct answers there are, they aren't sure. In some cases, you may be provided with the hint of selecting two or three out of five possible choices.

Point and click – this has a java-style interface. You will be presented with a picture, and then asked to pick a location on the picture to simulate a user selection. When you pick, a mark will be left on the image to indicate your selection. Each time you pick in a different area, the mark will shift to the new location. You do not have to pick an exact point…a general target area is all that is required.

Matching format – you are probably familiar with this style of question from elementary school. You will be presented with two columns. One column may have assorted terms and the second column the definitions. You then are expected to drag the terms to the correct definition to match the items.

What are the exercises like?

Each performance based question follows the same process. You will be asked to open a file. You may be asked to open a specific view or a sheet. This tests your ability to navigate around Revit, so be familiar with the Project Browser and how it works. You may be asked to orient the model to a specific orientation or level. You will then be required to perform a specific task, either adding or modifying an element or determining an element's properties. You will then be asked to fill an answer into an input box. The answer will be either text or numeric. The answer must match exactly with what is presented on screen, so it is best to Copy and Paste between Revit and the secure browser. If you miss a punctuation mark, use the wrong case, or spell something wrong, your answer will be marked wrong.

If you are unsure about a question in the exam and want to mark it for review, perform a SaveAs Copy of the file, so you don't lose any work you did.

Any tips?

I suggest you read every question at least twice. Some of the wording on the questions is tricky.

Be well rested and be sure to eat before the exam. Most testing centers do not allow food, but they may allow water. Keep in mind that you can take a break if you need one.

Relax. Maintain perspective. This is a test. It is not fatal. If you fail, you will not be the first person to have failed this exam. Failing does not mean you are a bad designer or architect or even a bad person. It just means you need to study the software more.

Have Revit open and ready to go before you start the test. Verify that you know where the drawing files are located so you don't have to search for them every time you need to open a file. Write down the file path on your scratch paper just in case you panic. Don't close the file. Occasionally, different questions will use the same file, so if you already have the file open that saves time. You will not open and use every file in the data set, so do not open every file in anticipation of using it. Practice switching windows before the exam. Open Internet Explorer and open a session of Revit. Then use Alt-Tab to switch between the windows. Practice this until you are comfortable.

Remember to write down your login name and password for your account. The proctor will not be able to help you if you forget.

Practice Exams

I have created a user practice exam and a professional practice exam. Do not memorize the answers. The questions on the practice exams will not be the same as the questions on the actual exam, but they may be similar. For example, the User practice exam will ask you to place stairs with specific properties and the Certified User exam will also ask you to place stairs with specific properties – but the properties will be different and the size and location of the stairs will be different. In the Professional practice exam, you will be asked to use a linked file and use Copy/Monitor and then in the Certified Professional practice exam, you will be asked to perform the same task. However, the files will be different and what you will be monitoring will be different.

Building Information Modeling and Revit Basics

This lesson addresses the following certification exam questions:

- Building Information Modeling
- User Interface
- Building Elements
- Revit Projects

There will be at least one question on the certification exam regarding Building Information Modeling. You will be expected to understand what BIM means and how it works. Autodesk is extremely proud that Revit is BIM software.

BIM means that Revit uses intelligent objects to create and manage a building model. In AutoCAD, you draw a set of lines to symbolize a door. In Revit, you place a door object which has parameters embedded in the object. These parameters contain data concerning the door: everything from the material, cost, and size to function and manufacturer information. This information can be leveraged to be used in schedules and in Excel spreadsheets. You can create an unlimited number of views for your building model and they all reside in a single file.

Revit boasts "bidirectional associativity," which means that if you make a change in one view, all related views also update.

Revit has parametric relationships within the model. For example, floors are constrained to walls, so if a wall is shifted in any direction, the floor will automatically update.

When you first launch Revit, a startup window named Recent Files is displayed.

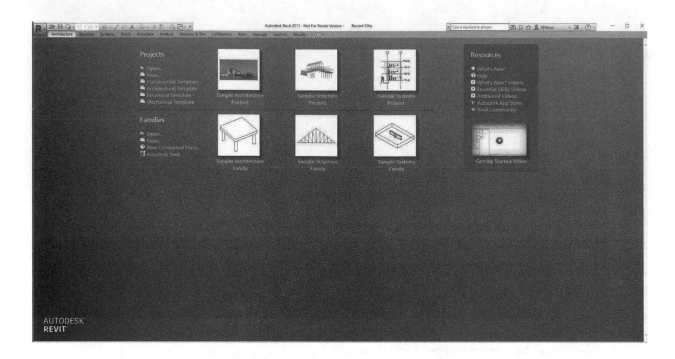

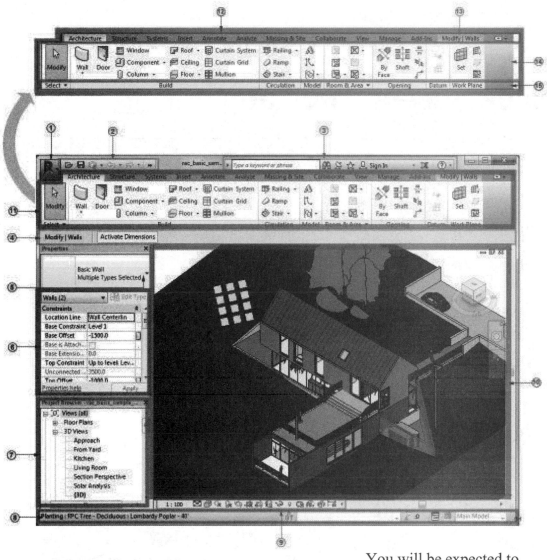

1	Application Menu
2	Quick Access Toolbar
3	InfoCenter
4	Options Bar
5	Type Selector
6	Properties Palette
7	Project Browser
8	Status Bar
9	View Control Bar
10	Drawing Area
11	Ribbon
12	Tabs on the ribbon
13	A contextual tab on the ribbon, providing tools relevant to the selected object or current action
14	Tools on the current tab of the ribbon
15	Panels on the ribbon

You will be expected to identify the different areas of the Revit User Interface in the exam.

For example, you may have a question asking you to indicate where the View Control Bar is located.

Exercise 1-1

Quick Access Toolbar

Drawing Name: **(none, start from scratch)**
Estimated Time to Completion: 10 Minutes

Scope
Learn how to add and remove tools from the Quick Access Toolbar.

Solution

1.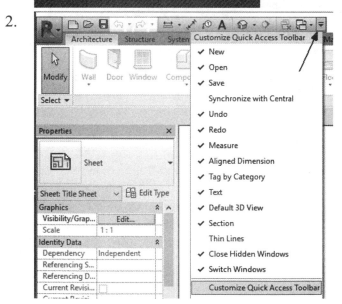

    Select the **Architectural Template** under Projects to start a new project.

2.

    Select the drop-down arrow on the Quick Access toolbar.
    Enable **New**.
    Disable **Synchronize with Central**.
    Disable **Thin Lines**.

The Quick Access toolbar updates with the new settings.

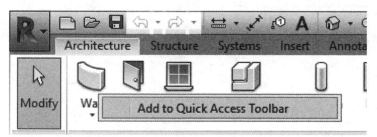

3. Place your mouse over the Wall tool on the Architecture ribbon.
Right click and select **Add to Quick Access Toolbar**.

Note: The Wall tool is grayed out unless you are in a plan view.

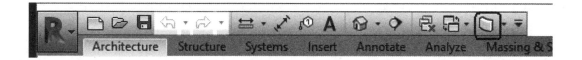

The Wall tool is added to the Quick Access toolbar (QAT).

4. Select the **Wall** tool on the QAT and place a wall in the drawing area.
To place the wall, just select two points like you are drawing a line.
Then press ESC to release the command.
Select the Wall and note that the ribbon changes to Modify mode.

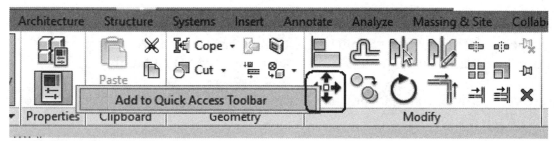

5. Activate the **Modify** ribbon.
Right click on the **Move** tool.
Select **Add to Quick Access Toolbar**.

The Quick Access toolbar now displays the Move tool.

6. Left click anywhere in the drawing area. This releases the wall from selection.

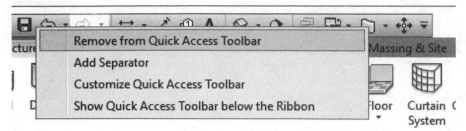

7. Right click on the **Wall** tool on the Quick Access Toolbar.
Select **Remove from Quick Access Ribbon**.

8. The Wall tool is removed.

9. Close the project by pressing **Ctrl+W**. When prompted if you want to save, press **No**.

The Quick Access toolbar behaves like the ribbon as some tools may become disabled depending on the mode you are in.

Building Elements are used to create a building design. There are five classes of building elements: host, component, datum, annotation, and view. Building elements fall into three categories: Model, View, and Annotation. To pass the User exam, users need to identify which category a building element falls in.

Each element falls into a category, such as wall, column, door, window, furniture, etc. Each category contains different families. Each family can have more than one type. The type is usually determined by the size or parameters assigned to that family.

These are very difficult concepts for many students, especially if they have been used to dealing with lines, circles, and arcs.

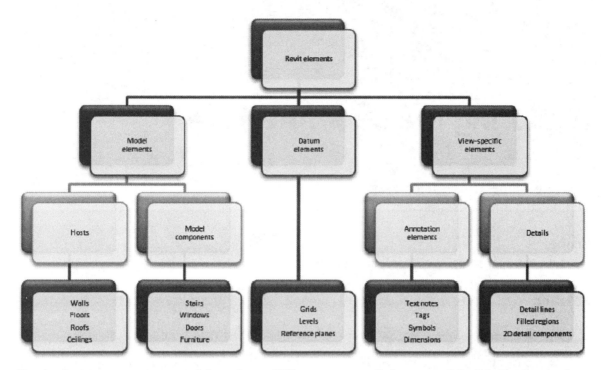

Revit elements are separated into three different types of elements: Model, Datum and View-specific. Users are expected to know if an element is model, datum or view-specific.

Model elements are broken down into categories. A category might be a wall, window, door, or floor. If you look in the Project Browser, you will see a category called Families. If you expand the category, you will see the families for each category in the current project. Each family may contain multiple types.

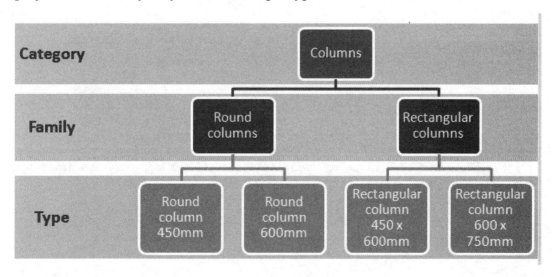

Every Revit file is considered a Project. A Revit project consists of the Project Environment, components, and views. The Project Environment is managed in the Project Browser.

Exercise 1-2

Exploring the User Interface

Drawing Name: c_user interface.rvt
Estimated Time to Completion: 5 Minutes

Scope
Review the user interface to prepare for the exam.

Solution

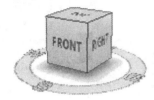

1. The file will open in a 3D view. Note that there is a ViewCube in the upper left corner.

- Views (all)
  - Floor Plans
    - Basement
    - High Roof
    - **Level 1**
    - Low Roof
    - Parapet
    - Site
    - T.O. Footing

2. Open the Level 1 Floor Plan view.

Double left click on Level 1 listed in the Project Browser.

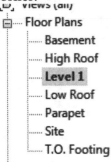

The ViewCube is only visible in 3D views. This is a possible question on the exam.

Note that the ViewCube is no longer visible and has been replaced with the Navigation Bar.

Exercise 1-3

Recover and Use Backup Files

Drawing Name: **new**
Estimated Time to Completion: 15 Minutes

Scope
Recover and Use Backup Files

Solution

1. Close any open projects.
   Go to the Application Menu and select **Close** or press **Ctrl+W** on the keyboard.

2.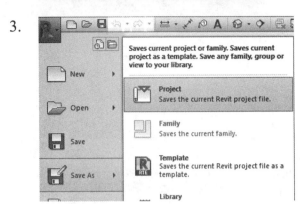

Select **Architectural Template** under Projects.
This starts a new project using the Architectural template.

3.

Go to **File→Save As→ Project**.

Saves current project or family. Saves current project as a template. Save any family, group or view to your library.

New

Open

Save

Save As

Project
Saves the current Revit project file.

Family
Saves the current family.

Template
Saves the current Revit project file as a template.

Library

4.

Options...

Select the **Options** button next to the file name.

5.

File Save Options

Maximum backups: 5

Set the Maximum backups: to **5**.

Some students prefer not to save any backups so that their flash drive doesn't fill up. Those students set the number of backups to 0.

Notice that the number of backups is unique to each project. This might be a question on the exam.

Press **OK**.

6.

File name: ex1-3

Files of type: Project Files (*.rvt)

Save as *ex1-3.rvt*.

7.   Draw four walls.

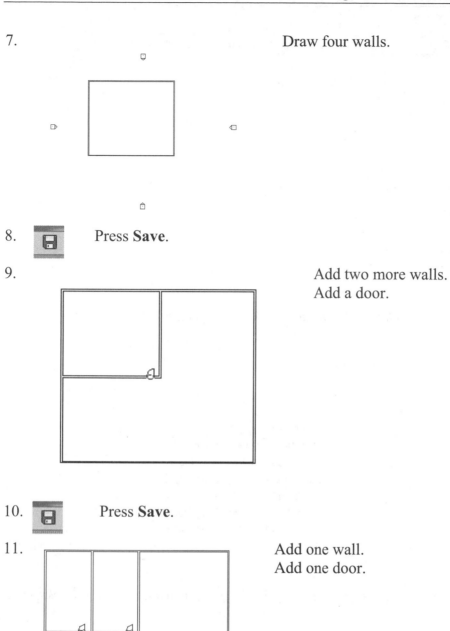

8.   Press **Save**.

9.   Add two more walls.
                                      Add a door.

10.   Press **Save**.

11.   Add one wall.
                                      Add one door.

12.   Press **Save**.

13. 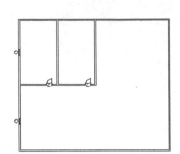 Add two windows.

14. Press **Save**.

15. Select **Open**.

16. ex1-3.0001
    ex1-3.0002
    ex1-3.0003
    ex1-3.0004
    ex1-3

Note that you have several versions of ex1-3.

The .0000x indicates the backup number.

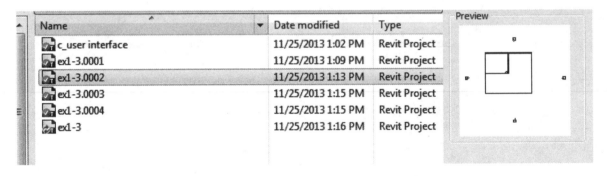

Name	Date modified	Type
c_user interface	11/25/2013 1:02 PM	Revit Project
ex1-3.0001	11/25/2013 1:09 PM	Revit Project
ex1-3.0002	11/25/2013 1:13 PM	Revit Project
ex1-3.0003	11/25/2013 1:15 PM	Revit Project
ex1-3.0004	11/25/2013 1:15 PM	Revit Project
ex1-3	11/25/2013 1:16 PM	Revit Project

Note that you can highlight a version and check in the preview window which backup you want to select.

17. Open *ex1-3.0001.rvt.*
This is the first save you did.

ex1-3.0001 - Floor Plan: Level 1

18. *Note the file name at the top of the screen.*
Close all files without saving.

Exercise 1-4
Design Options

Drawing Name: **i_Design_Options**
Estimated Time to Completion: 90 Minutes

Scope
Use of Design Options

Solution

1. 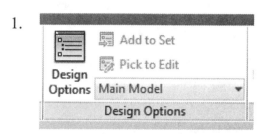 Activate the **Manage** ribbon.
Select **Design Options** under the Design Options panel.

2. 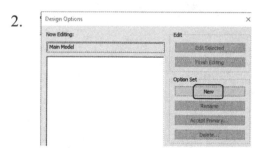 Select **New** under Option Set.
Select **New** a second time.

3. 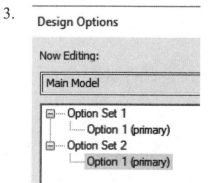 There should be two Option Sets displayed in the left panel.
Each Option set represents a design choice group.
The Option set can have as many options as needed.
The more options, the larger your file size will become.

4.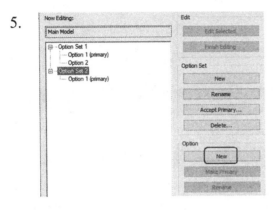

Highlight the **Option Set 1**.
Select the **New** button under Option. Note that Option Set 1 now has two sub-options.

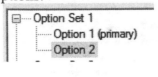

5.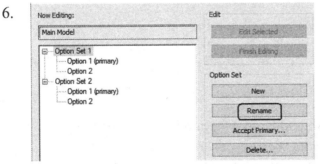

Highlight the **Option Set 2**.
Select the **New** button under Option. Note that Option Set 2 now has two sub-options.

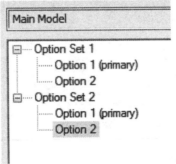

6.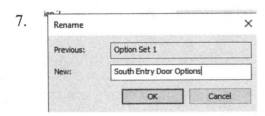

Highlight **Option Set 1**.
Select **Rename**.

7.

Rename Option Set 1 **South Entry Door Options**.
Press **OK**.

Rename	×
Previous:	Option Set 1
New:	South Entry Door Options
	OK Cancel

8.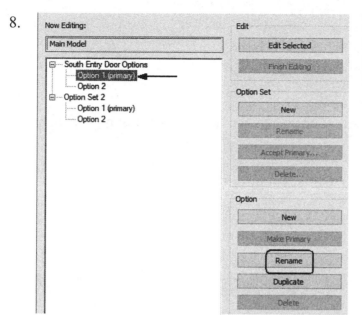

Highlight **Option 1 (primary)** under the South Entry Door Options. Select **Rename**.

9.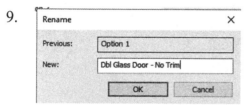

Rename to **Dbl Glass Door - No Trim**. Press **OK**.

10.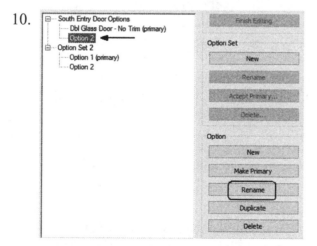

Highlight **Option 2** under the South Entry Door Options. Select **Rename**.

11.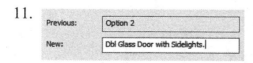

Rename to **Dbl Glass Door with Sidelights**. Press **OK**.

12. Highlight **Option Set 2**.
Select **Rename**.

13. Rename Option Set 2 **Office Layout Design Options**.
Press **OK**.

14. Highlight **Option 1 (primary)** under the **Office Layout Design Options**.
Select **Rename**.

15. Rename to **Indented Walls**.
Press **OK**.

16. 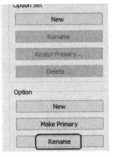 Highlight **Option 2**.
Select **Rename**.

17. Rename Option 2 **Flush Walls**.
Press **OK**.

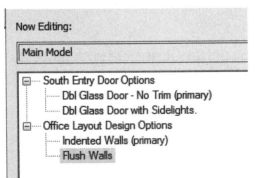 You should have two Option Sets.
Each Option Set should have two options.

An Option Set can have as many options as you like, but the more option sets and options, the larger your file size and the more difficult it becomes to manage.

18. Close the Design Options dialog.

19. 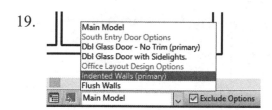   Note in the bottom of the window, you can select which Option set you want active.

20. Using **Duplicate View→Duplicate**, create four copies of the Level 1 view.

Rename the duplicate views:
Level 1 - Office Layout Indented Walls
Level 1 - Office Layout Flush Walls
Level 1 - South Entry Dbl Glass Door – No Trim
Level 1 - South Entry Dbl Glass Door with Sidelights

To rename, highlight the level name and press F2.

21. 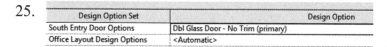   Using **Duplicate View→Duplicate**, create two copies of the South Elevation view.
Rename the duplicate views:
South Entry Dbl Glass Door – No Trim
South Entry Dbl Glass Door with Sidelights

22.    Activate **Level 1 - South Entry Dbl Glass Door – No Trim**.

23. 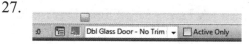   In the Properties pane:
Select Edit for **Visibility/Graphics Overrrides**.

24.    Select the **Design Options** tab.

25.

Design Option Set	Design Option
South Entry Door Options	Dbl Glass Door - No Trim (primary)
Office Layout Design Options	<Automatic>

Set South Entry Door Options to **Dbl Glass Door - No Trim (primary)**.
Press **OK**.

26.    Set the Design Option to **Dbl Glass Door - No Trim (primary)**.

27. 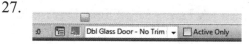   Uncheck **Active Only**.

28.

Select the south horizontal wall.

29.

Activate the **Manage** ribbon.
Under Design Options, select **Add to Set**.
The selected wall is added to the **Dbl Glass Door - No Trim (primary)** *set.*
We need to add the wall to the set so we can place a door. Remember doors are wall-hosted.

30.

Activate the **Architecture** ribbon.
Select the **Door** tool from the Build panel.

31.

Set the Door type to **Dbl-Glass 1: 68″ x 84″**.

32.

Place the door as shown.

33.

Activate **Level 1 - South Entry Dbl Glass Door with Sidelights**.

34.

In the Properties pane:
Select **Edit** Visibilities/Graphics Overrides.

35. Activate the **Design Options** tab.

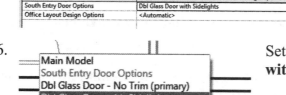

Set **Dbl Glass Door with Sidelights** on South Entry Door Options. Press **OK**.

36.

Main Model
South Entry Door Options
Dbl Glass Door - No Trim (primary)
Dbl Glass Door with Sidelights
Office Layout Design Options
Indented Walls (primary)
Flush Walls

Dbl Glass Door with Sideli ▾ ☑ Active Only

Set the Design Option to **Dbl Glass Door with Sidelights**.

37.

Architecture   Structure

Wall Door Window

Activate the **Architecture** ribbon.
Select the **Door** tool from the Build panel.

38.

Place a **Double-Raised Panel with Sidelights: 68″ x 80″** door as shown.

Properties

Double-Raised Panel with Sidelights
68" x 80"

39.

Elevations (Elevation 1)
    East
    North
    South
    South Entry Dbl Glass Door with Sidelights
    South Entry Dbl Glass Door – No Trim
    West

Activate the **South Entry Dbl Glass Door – No Trim** elevation.

40.

Parts Visibility	Show Original
Visibility/Graphics Overrides	Edit...
Graphic Display Options	Edit...
Underlay	None

In the Properties pane:
Select **Edit** Visibilities/Graphics Overrides.

41. Activate the **Design Options** tab.

Design Option Set	Design Option
South Entry Door Options	Dbl Glass Door - No Trim (primary)
Office Layout Design Options	<Automatic>

Set **Dbl Glass Door - No Trim** on South Entry Door Options. Press **OK**.

42.

Elevations (Elevation 1)
    East
    North
    South
    South Entry Dbl Glass Door with Sidelights
    South Entry Dbl Glass Door – No Trim
    West

Activate the **South Entry Dbl Glass Door with Sidelights** elevation.

43.

Parts Visibility	Show Original
Visibility/Graphics Overrides	Edit...
Graphic Display Options	Edit...
Underlay	None

In the Properties pane:
Select **Edit** Visibilities/Graphics Overrides.

44. Activate the **Design Options** tab.

Set **Dbl Glass Door with Sidelights** on South Entry Door Options.

Design Option Set		Design Option
South Entry Door Options	Dbl Glass Door with Sidelights	
Office Layout Design Options	<Automatic>	

Press **OK**.

45. ⊟ 📖 Sheets (all)
    ⊢—— A101 - South Entry Door Option
    └—— A102 - Office Layout Options

Activate the Sheet named **South Entry Door Options**.

46.

Drag and drop the two South Entry Option elevation views on the sheet.

47. Switch to 3D view.

48. Use **Duplicate View→Duplicate** to create two new 3D views.

⊟—— 3D Views
    ├—— 3D - South Entry Dbl Glass Door - No Trim
    ├—— 3D - South Entry Dbl Glass Door with Sidelights
    └—— {3D}

Rename the views:
3D - South Entry Dbl Glass Door - No Trim
3D - South Entry Dbl Glass Door with Sidelights

49. ⊟—— 3D Views
    ├—— **3D - South Entry Option 1**
    ├—— 3D - South Entry Option 2
    └—— {3D}

Activate **3D - South Entry Dbl Glass Door - No Trim**.

50.

Detail Level	Coarse
Visibility/Graphics Overrides	Edit...
Visual Style	Hidden Line
Graphic Display Options	Edit...

In the Properties pane:
Select **Edit** Visibilities/Graphics Overrides.

51. Activate the **Design Options** tab.

Design Option Set	
South Entry Door Options	Dbl Glass Door - No Trim (primary)
Office Layout Design Options	<Automatic>

Set **Dbl Glass Door - No Trim** on South Entry Door Options.
Press **OK**.

52. ⊟—— 3D Views
    ├—— 3D - South Entry Dbl Glass Door - No Trim
    ├—— **3D - South Entry Dbl Glass Door with Sidelights**
    └—— {3D}

Activate **3D - South Entry Dbl Glass Door with Sidelights**.

53.

Display Model	Normal
Detail Level	Coarse
Visibility/Graphics Overrides	Edit...
Visual Style	Hidden Line
Graphic Display Options	Edit...

In the Properties pane:
Select **Edit** Visibilities/Graphics Overrides.

54. Activate the **Design Options** tab.

Design Option Set	
South Entry Door Options	Dbl Glass Door with Sidelights
Office Layout Design Options	<Automatic>

Set **Dbl Glass Door with Sidelights** on South Entry Door Options.
Press **OK**.

55.

Sheets (all)
  A101 - South Entry Door Options
  A102 - Office Layout Options

Activate the Sheet named **South Entry Door Options**.

56.

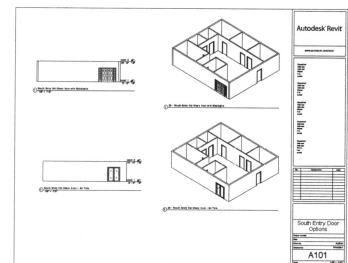

Drag and drop the 3D views onto the sheet.

57.

Floor Plans
  Level 1
  Level 1 - Office Layout Flush Walls
  Level 1 - South Entry Dbl Glass Door with Sidelights
  Level 1 - South Entry Dbl Glass Door – No Trim
  Level 1- Office Layout Indented Walls
  Level 2

Activate **Level 1 - Office Layout Indented Walls**.

58.

Detail Level	Coarse
Visibility/Graphics Overrides	Edit...
Visual Style	Hidden Line

In the Properties pane:
Select Edit for **Visibility/Graphics Overrrides**.

59. Design Options   Select the **Design Options** tab.

60.

Design Option Set	
South Entry Door Options	<Automatic>
Office Layout Design Options	Indented Walls (primary)

Set Office Layout Design Options to **Indented Walls (primary)**.
Press **OK**.

61. Indented Walls (primary)   Active Only

Set the Design Option to **Indented Walls (primary)**.

62. 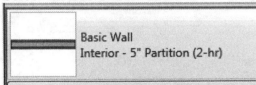 Select the Wall tool from the Architecture ribbon.
Select the Basic Wall: Interior - 5" Partition (2-hr).

63. 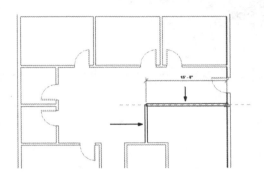 Place the two walls indicated.
The vertical wall is placed at the midpoint of the small horizontal wall to the south.
The horizontal wall is aligned with the wall indicated by the dashed line.

64.  Activate the **Architecture** ribbon.
Select the **Door** tool from the Build panel.

65. 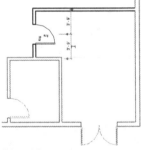 Place a **Sgl Flush 36″ x 80″** door as shown.

66.  Activate **Level 1 - Office Layout Flush Walls**.

67. In the Properties pane:
Select Edit for **Visibility/Graphics Overrrides**.

Detail Level	Coarse
Visibility/Graphics Overrides	Edit...
Visual Style	Hidden Line

68. Design Options   Select the **Design Options** tab.

69. Set Office Layout Design Options to **Flush Walls**.
Press **OK**.

Design Option Set	Design Option
South Entry Door Options	<Automatic>
Office Layout Design Options	Flush Walls

70. 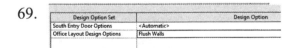 Set the Design Option to **Flush Walls**.

71. ☐ Active Only   Uncheck **Active Only**.

72. Activate the **Architecture** ribbon.
Select the **Wall** tool from the Build panel.

73. On the Properties pane:
Select the **Basic Wall: Interior - 5″ Partition (2 hr)** wall type.

74. Add the wall shown.

Note that the walls and door added for the Indented Walls option are not displayed.

75. Activate the **Architecture** ribbon.
Select the **Door** tool from the Build panel.

76. Place a **Sgl Flush 36″ x 84″** door as shown.

77. Activate the **Manage** ribbon.
Select **Design Options** on the Design Options panel.

78. Select **Finish Editing**.
Close the dialog.

79. Change the Design Option to **Main Model**.

80. Note that if you hover your mouse over the element, it will display which Option set it belongs to.

This only works if Active Only or Exclude Options is disabled.

(South Entry Door Options : Dbl Glass Door - No Trim) :
Doors : Dbl-Glass 1 : 68" x 84"

81. Activate the Sheet named **Office Layout Options**.

Sheets (all)
A101 - South Entry Door Options
A102 - Office Layout Options

82. Drag and drop the two Office Layout options onto the sheet.

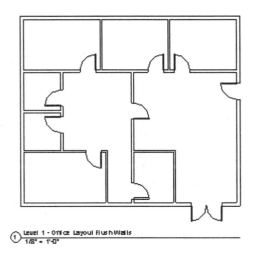

Level 1 - Office Layout Flush Walls
1/8" = 1'-0"

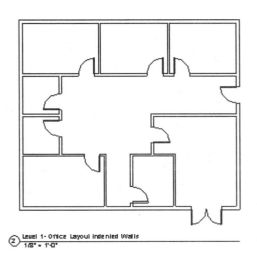

Level 1 - Office Layout Indented Walls
1/8" = 1'-0"

83. Activate the Manage ribbon.
Select **Design Options**.

Add to Set

Pick to Edit

Design Options   Main Model

Design Options

84. *Let's assume that the client decided they prefer the flush walls option.*

Highlight the **Flush Walls** option.

Now Editing:

Main Model

South Entry Door Options
Dbl Glass Door - No Trim (primary)
Dbl Glass Door with Sidelights
Office Layout Design Options
Indented Walls (primary)
Flush Walls

85. Select **Make Primary**.

Make Primary

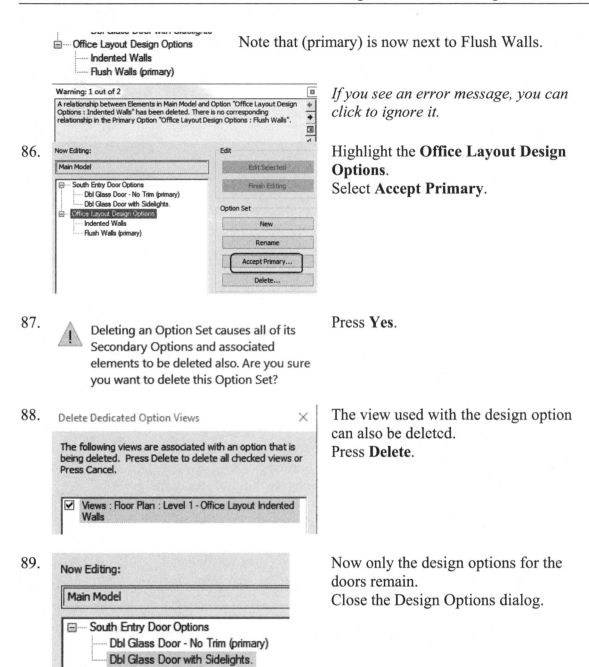

Note that (primary) is now next to Flush Walls.

If you see an error message, you can click to ignore it.

86. Highlight the **Office Layout Design Options**.
Select **Accept Primary**.

87. Press **Yes**.

88. The view used with the design option can also be deleted.
Press **Delete**.

89. Now only the design options for the doors remain.
Close the Design Options dialog.

90. Close without saving.

Exercise 1-5
Phases

Drawing Name: c_phasing.rvt
Estimated Time to Completion: 75 Minutes

Scope
Properties
Filter
Phases
Rename View
Copy View
Graphic Settings for Phases

Solution

1. Select the **Open** tool.

2. File name: c_Phasing.rvt   Locate the *c_phasing.rvt* file. Select **Open**.

3. 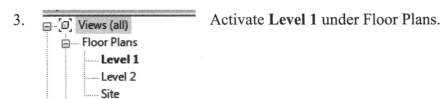   Activate **Level 1** under Floor Plans.

4. 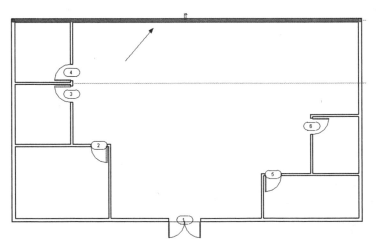   Select the wall indicated.
It should highlight.

5. 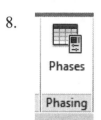 Scroll down to the Phasing category in the Properties panel on the upper left.

6. This wall was created in the New Construction Phase. Note that it is not set to be demolished.

7. Right click and press **Cancel** to deselect the wall. `Cancel`

8. Go to the **Manage** ribbon.
   Select **Phasing→Phases**.

9. 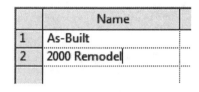 Rename Existing to **As-Built**.

10. Rename New Construction to **2000 Remodel**.

11. Highlight the **2000 Remodel**.
    Select **After**.

12. 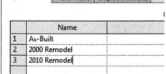 Name the new phase **2010 Remodel**.

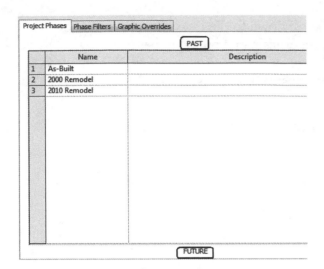

Note that the top indicates the past and the bottom indicates the future to help orient the phases.

13. Select the **Graphic Overrides** tab.

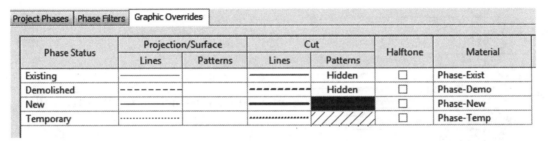

14. Note that in the Lines column for the Existing Phase, the line color is set to gray.

15.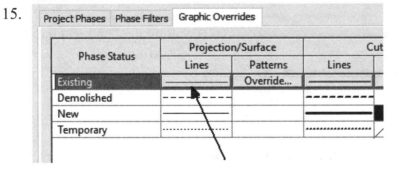

Highlight **Existing**. Click in the **Lines** column and the Line Graphics dialog will display.

Projection/Surface is what is displayed in the floor plan views.
Cut is the display for elevation or section views.
Override indicates you have changed the display from the default settings.

16. Set the Color to **Green** for the Existing phase by selecting the color button. Press **OK**.

Set the Color to **Blue** for the Demolished phase.
Set the Color to **Magenta** for the New phase.
Change the colors for both Projection/Surface and Cut.

Phase Status	Projection/Surface		Cut	
	Lines	Patterns	Lines	Patterns
Existing	———————		———————	Hidden
Demolished	– – – – – – –		– – – – – – –	Hidden
New	———————	Override...	———————	
Temporary	·················		·················	/////

17. | Project Phases | Phase Filters | Graphic Overrides | Select the **Phase Filters** tab.

18.

	Filter Name	New	Existing	Demolished	Temporary
1	Show All	By Category	Overridden	Overridden	Overridden
2	Show Demo + New	By Category	Not Displayed	Overridden	Overridden
3	Show Previous + Dem	Not Displayed	Overridden	Overridden	Not Displayed
4	Show Previous + New	By Category	Overridden	Not Displayed	Not Displayed
5	Show Previous Phase	Not Displayed	Overridden	Not Displayed	Not Displayed

Note that there are already phase filters pre-defined that will control what is displayed in a view.

19.   New   Press the **New** button on the bottom of the dialog.

20.

	Filter Name
1	Show All
2	Show Demo + New
3	Show Previous + Dem
4	Show Previous + New
5	Show Previous Phase
6	Show Existing

Change the name for the new phase filter to **Show Existing**.
Show Previous + Demo will display existing plus demo elements, but not new.
Show Previous + New will display existing plus new elements, but not demolished elements.

21. In the New column, select **Overridden**.
In the Existing column, select **Overridden**.
This means that the default display settings will use the new color assigned.
In the Demolished column, select **Not Displayed**.

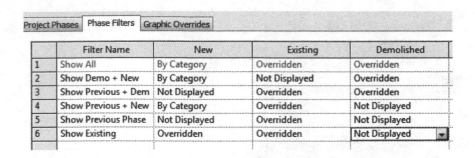

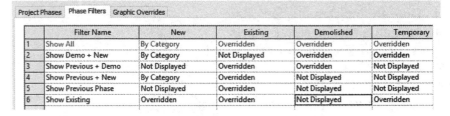

22. Use Overridden to display the colors you assigned to the different phases.
    Verify that in the Show Previous + Demo phase New elements are not displayed.
    Verify that in the Show Previous + New phase Demolished elements are not displayed.
    Press **Apply** and **OK** to close the Phases dialog.

23. ![Filter icon] Window around the entire floor plan.
    Select the **Filter** button.

    Filter

24. | Category: | Count: |
    |---|---|
    | ☐ Door Tags | 6 |
    | ☑ Doors | 6 |
    | ☑ Floors | 1 |
    | ☑ Walls | 12 |

    Uncheck **Door Tags**.
    Tags and annotations are not affected by phases.
    Press **OK.**

25. | Phasing | |
    |---|---|
    | Phase Created | As-Built |
    | Phase Demolished | None |

    Set the Phase Created to **As-Built**.

26.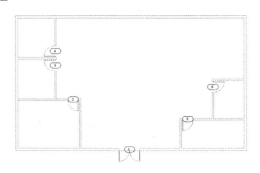

    Note that the view changes to display in Green.
    This is because we set the color Green to denote existing elements.

27.
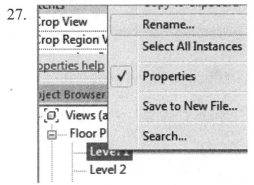

Next we create three Level 1 floor plan views for each phase.
Highlight **Level 1** under Floor Plan.
Right click and select **Rename**.

28.
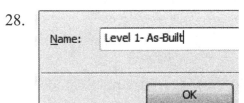

Rename the view **Level 1- As Built**.
Press **OK**.

29. Would you like to rename corresponding level and views?

Press **No**.

Yes   No

30.
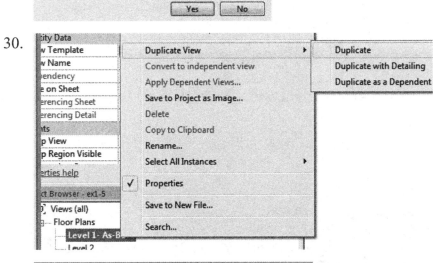

Highlight **Level 1- As Built** under Floor Plan.
Right click and select **Duplicate View→Duplicate**.

31.
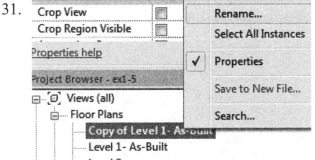

Highlight **Copy of Level 1-Existing** under Floor Plan.
Right click and select **Rename**.

32.

In the text field, enter **Level 1-2000 Remodel Demo**.
Press **OK**.

33.  Highlight **Level 1-As Built** under Floor Plan.

34. Right click and select **Duplicate View→Duplicate**.

35. 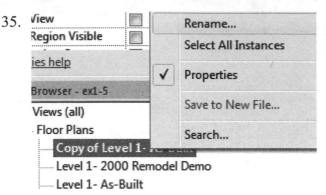 Highlight **Level 1-As Built Copy 1** under Floor Plan.
Right click and select **Rename**.

36. 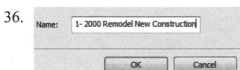 In the text field, enter **Level 1-2000 Remodel New Construction**.
Press **OK**.

37. 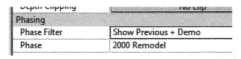 You should have three floor plan views listed:
As Built
2000 Demo
2000 New Construction.

38. 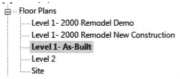 Activate the **Level 1-2000 Remodel Demo** view.

39. In the Properties dialog:

Depth Clipping	No clip
Phasing	
Phase Filter	Show Previous + Demo
Phase	2000 Remodel

Set the Phase to **2000 Remodel**.

Set the Phase Filter to **Show Previous + Demo**.
The previous phase to demo is As-Built. This means the view will display elements created in the existing and demolished phase.

40. Activate the **Level 1-As Built** view.

Floor Plans
- Level 1- 2000 Remodel Demo
- Level 1- 2000 Remodel New Construction
- **Level 1- As-Built**
- Level 2
- Site

The display does not show the graphic overrides. By default, Revit only allows you to assign graphic overrides to phases AFTER the initial phase. Because the As-Built view is the first phase in the process, no graphic overrides are allowed. The only work-around is to create an initial phase with no graphic overrides and go from there.

41. In the Properties dialog:

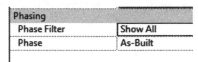

Set the Phase Filter to **Show All**.
Set the Phase to **As-Built**.

42.

Activate the **Level 1-2000 Remodel New Construction** view.

43. In the Properties dialog:

Set the Phase Filter to **Show Previous + New**.
This will display elements created in the Existing Phase and the New Phase, but not the Demo phase.
Set the Phase to **2000 Remodel**.

44.

Activate the **Level 1 - 2000 Remodel Demo** view.

45.

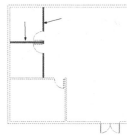

Hold down the Ctrl button.
Select the two walls indicated.

46. In the Properties pane:

Scroll down to the bottom.
In the Phase Demolished drop-down list, select **2000 Remodel**.

47. Press **OK**.

48. The demolished walls change appearance based on the graphic overrides.
Release the selected walls using right click→Cancel or by pressing ESCAPE.

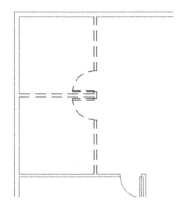

49.

Activate the **Modify** ribbon.
Use the **Demolish** tool on the Geometry panel to demolish the walls indicated.

50.

Note that the doors will automatically be demolished along with the walls. If there were windows placed, these would also be demolished. That is because those elements are considered *wall-hosted*.

Right click and select Cancel to exit the Demolish mode.

51.

This is how the Level 1- 2000 Remodel Demo view should appear.

If it doesn't, check the walls to verify that they are set to Phase Created: As Built, Phase Demolished: 2000 Remodel.

Phasing	
Phase Created	As-Built
Phase Demolished	2000 Remodel

52.

Activate the **Level 1 - 2000 Remodel New Construction** view.

53.

Select the **Wall** tool from the Architecture ribbon.

54.

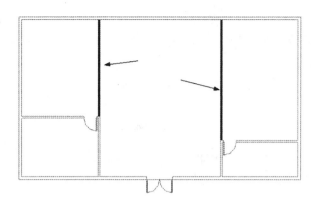

Place two walls as shown. Select the end points of the existing walls and simply draw up.
Right click and select **Cancel** to exit the Draw Wall mode.

55.

Select the **Door** tool under the Build panel on the Architecture ribbon.

56.

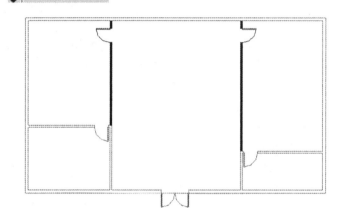

Place two doors as shown. Set the doors 3′ 6″ from the top horizontal wall. Flip the orientation of the doors if needed.

You can press the space bar to orient the doors before you left click to place.

Note that the new doors and walls are a different color than the existing walls.

57.

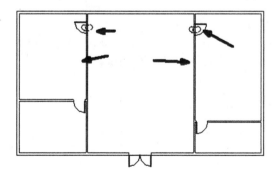

Select the doors and windows you just placed.
You can select by holding down the CONTROL key or by windowing around the area.

Note: If Door Tags are selected, you will not be able to access Phases in the Properties dialog.

58. Look in the Properties panel and scroll down to Phasing.

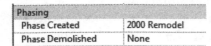

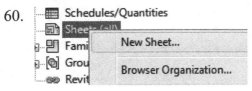   Note that the elements are already set to **2000 Remodel** in the Phase Created field.

59. Switch between the three views to see how they display differently.

60. 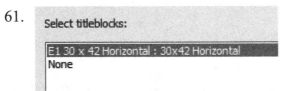   Highlight **Sheets** in the Project Browser. Right click and select **New Sheet**.

61.   Press **OK** to accept the default title block.

Select titleblocks:

E1 30 x 42 Horizontal : 30x42 Horizontal
None

62.    A view opens with the new sheet.

63.    Highlight the Level 1 - As Built Floor plan. Hold down the left mouse button and drag the view onto the sheet. Release the left mouse button to click to place.

64. A preview will appear on your cursor. Left click to place the view on the sheet.

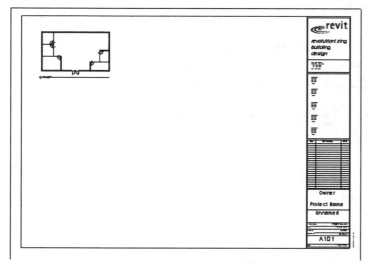

65. Highlight the Level 1 - 2000 Remodel Demo Floor plan.
Hold down the left mouse button and drag the view onto the sheet. Release the left mouse button to click to place.

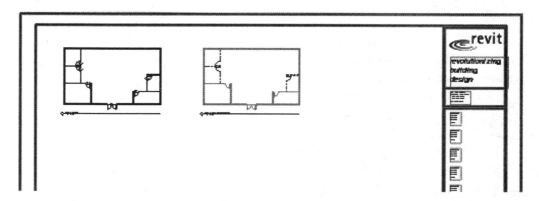

The two views appear on the sheet.

66. Highlight the Level 1 - 2000 New Construction plan.
Hold down the left mouse button and drag the view onto the sheet. Release the left mouse button to click to place.

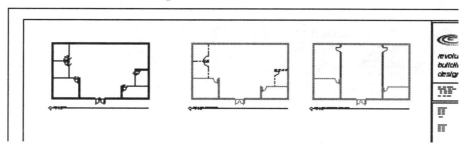

67. Zoom in to inspect the views.
    If the new walls do not display in magenta, go back to the Phase Filters and verify
    that New is set to **Overridden**.

	Filter Name	New
1	Show All	By Category
2	Show Demo + New	By Category
3	Show Existing	Overridden
4	Show Previous + Demo	Not Displayed
5	Show Previous + New	Overridden
6	Show Previous Phase	Not Displayed

Project Phases | **Phase Filters** | Graphic Overrides

68. Save as *ex1-5.rvt*.

Challenge Exercise:

Create two more views called Level 1 2010 Remodel Demo and Level 1 2010 Remodel
New Construction.

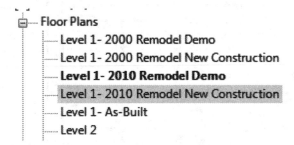

Floor Plans
- Level 1- 2000 Remodel Demo
- Level 1- 2000 Remodel New Construction
- **Level 1- 2010 Remodel Demo**
- Level 1- 2010 Remodel New Construction
- Level 1- As-Built
- Level 2

Set the Phases and phase filters to the new views.

The 2010 Remodel Demo view should be set to:

Phasing	
Phase Filter	Show Previous + Demo
Phase	2010 Remodel

The 2010 Remodel New Construction view should be set to:

Phasing	
Phase Filter	Show Previous + New
Phase	2010 Remodel

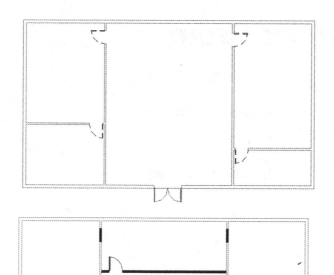

On the 2010 Remodel Demo view:
Demo all the interior doors.

For the 2010 remodel new
construction, add the walls and
doors as shown.

Note you will need to fill in the
walls where the doors used to be.

Add the 2010 views to your sheet.

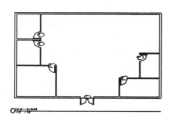
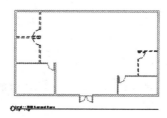
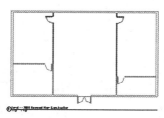

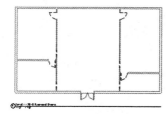
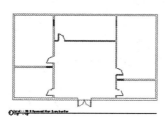

Answer this question:

When should you use phasing as opposed to design options?

Certified User Practice Exam

1. Select the answer which is NOT an example of bidirectional associativity:
   - A. Flip a section line and all views update.
   - B. Draw a wall in plan view and it appears in all other views.
   - C. Change an element type in a schedule and the change is displayed in the floor plan view as well.
   - D. Flip a door orientation so the door swing is on the exterior of the building.

2. Select the answer which is NOT an example of a parametric relationship:
   - A. A floor is attached to enclosing walls. When a wall moves, the floor updates so it remains connected to the walls.
   - B. A series of windows are placed along a wall using an EQ dimension. The length of the wall is modified and the windows remain equally spaced.
   - C. A door is placed in a wall. The wall is moved and the door remains constrained in the wall.
   - D. A shared parameter file is loaded to the server.

3. Which tab does NOT appear on Revit's ribbon?
   - A. Architecture
   - B. Basics
   - C. Insert
   - D. View

4. Which item does NOT appear in the Project Browser?
   - A. Families
   - B. Groups
   - C. Callouts
   - D. Notes

5. Which is the most recently saved backup file?
   - A. office.0001
   - B. office.0002
   - C. office.0003
   - D. office.0004

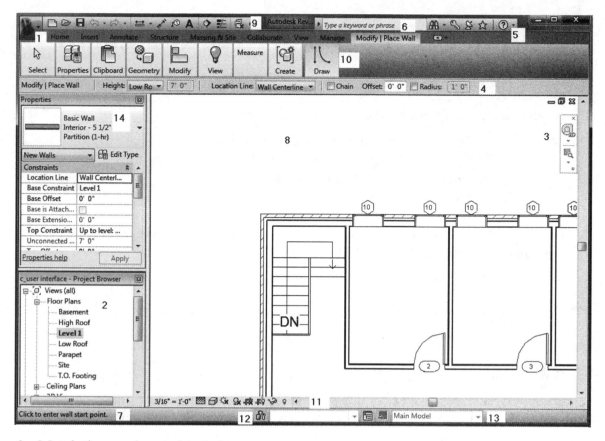

6. Match the numbers with their names.

View Control Bar	InfoCenter
Project Browser	Status Bar
Navigation Bar	Properties Pane
Options Bar	Application Menu
Design Options	Drawing Area
Help	Quick Access Toolbar
Ribbon	Worksets

Certified Professional Practice Exam

1. When using design options, the active option is the _____
   - A. Preferred design option in the design option set.
   - B. Part of the building that is not defined using design options.
   - C. Design options currently being edited.
   - D. Collection of all design options.

2. A _____ is a rule that you apply to a view to control the display of elements based on the phase status.
   - A. View Template
   - B. Display State
   - C. Phase Filter
   - D. Design Option

3. If you demolish an element in one view:
   - A. It is displayed as demolished in all views with the same phase.
   - B. It is displayed as demolished only in plan and elevation views with the same phase.
   - C. It is displayed as demolished in all views with the same phase except for section views.
   - D. It is displayed as demolished in that view only.

4. When a view is opened or created, by default the View Phase is set to:
   - A. Default
   - B. Demolished
   - C. Existing
   - D. New Construction

5. The two properties used to control the phase and display of a view are:
   - A. Phase Filter
   - B. Phase
   - C. Graphic Display Options
   - D. Visibility/Graphics Overrides

6. The _____ is the entire building model, excluding any design options.

    A. The Main Model
    B. The Basic Model
    C. The Primary Model
    D. The Primary Option

7. The _____ is the preferred option in a Design Option set.

    A. Main Model
    B. Primary Option
    C. Secondary Option
    D. Active Option

8. The _____ is the design option which is active and currently being edited.

    A. Main Model
    B. Primary Option
    C. Active Option
    D. Default Option

9. A _____ is a view that is dedicated to a specific design option. When the view is active, Revit displays the design option along with the rest of the building model.

    A. Dedicated view
    B. Phased view
    C. Design Option view
    D. Primary Option view
    E. Active view

Answers:
    1) C; 2) C; 3) A; 4) D; 5) A & B; 6) A; 7) B; 8) C; 9) A

The Basics of Building a Model

This lesson addresses the following types of problems:

- Wall Properties
- Compound Walls
- Stacked Walls
- Doors and Windows
- Using the Array Tools
- In-Place Mass

In the certification exam, most of the wall problems follow these steps:

- Place a wall of a specific element type. (Be able to select wall type.)
- Place a wall by setting the location line. (Understand how to use the location line setting.)
- Place a wall using different Option Settings. (Understand how to use the Options Settings when placing a wall.)
- After placing the wall, place a dimension to determine if the wall was placed correctly.
- After placing the wall, inspect the element properties to determine if the wall was placed correctly.

Users will need to be familiar with the different parameters in walls and compound walls. The user should also know which options are applied to walls and when those options are available.

Walls are system families. They are project-specific. This means the wall definition is only available in the active project. However, there are methods to copy a wall definition from one project to another. This will also be covered in this lesson.

Exercise 2-1
Wall Options

Drawing Name: **i_firestation_basic_plan.rvt**
Estimated Time to Completion: 10 Minutes

Scope
Exploring the different wall options

Solution

1. Floor Plans
    Ground Floor
    Lower Roof
    Main Floor
    Main Roof
    Site
    T.O. Footing
    T.O. Parapet

    Activate the **Ground Floor** floor plan.

2. Zoom into the area where the green polygon is.

3. Select **Wall** from the Architecture ribbon.

4. Set the Wall Type to **Generic – 6″** in the Properties pane.

5. Set the Location Line to **Core Face:Exterior**.

6. Select the **Rectangle** tool on the Draw panel.

7.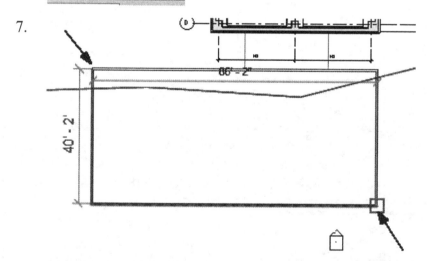
    Select the two points indicated to place the rectangle.

8. Select the **Line** tool from the Draw panel.

9.
    Start the line at the midpoint of the lower horizontal wall.

10.

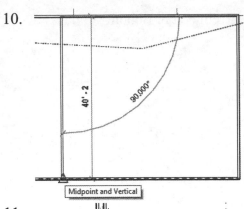

Bring the line end up to the midpoint of the upper horizontal wall.
Left click to finish placing the wall.
Exit the wall tool.

11.

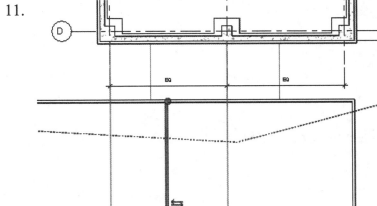

Select the vertical wall.
Two temporary (listening) dimensions will appear.

Change the right dimension to **12′ [3600 mm]**.

12.

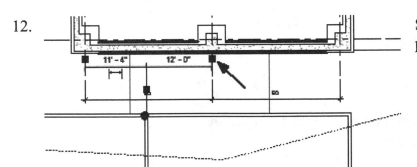

Select the witness grip point indicated.

13.

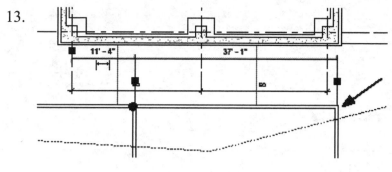

Move the witness line to the right vertical wall.

Note that the dimension updates.

The dimension should display 37' 1". This is the answer you would enter in the certification exam. If you did not get that answer, check your location line setting.

14. Close the file without saving.

Exercise 2-2
Array Components

Drawing Name: **i_array.rvt**
Estimated Time to Completion: 30 Minutes

Scope
Create a Radial array of a component
Create a Linear array of a component

In the Certification exam, you will be asked to place an array and then take a measurement to verify the array was placed correctly.

Solution

1. Open *i_array.rvt*.

2. Floor Plans
    — Ground Floor
    — Lower Roof
    — Main Floor
    — **Main Floor – Kitchen**
    — Main Roof
    — Site
    — T. O. Footing
    — T. O. Parapet

    Activate the **Main Floor – Kitchen** floor plan.

    This is a dependent view.

3.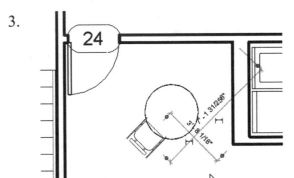

    Select the Chair- Stacking located at the table near the door.

4. Select the **Array** tool from the Modify panel.

    Array (AR)

    Modify

5. ☑ Group and Associate

    Set the Array type to **Radial**.

6. 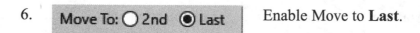 Enable Move to **Last**.

7.  Set the Number to **3**.

8. Select the **Place** button next to Center of Rotation.

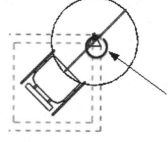

9. 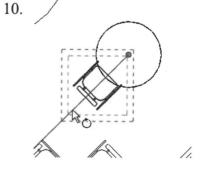 Select the midpoint of the table.

10. Select the midline of the Chair-Stacking as the starting angle.

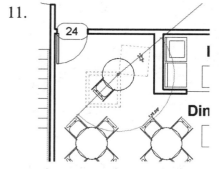

11. Sweep the cursor in the counter clockwise direction until you see an angle of 175.
Left click.

12. You will see a preview with a number 3.
Press ENTER to accept.
Left click anywhere in the window to release the selection and end the command.

13. 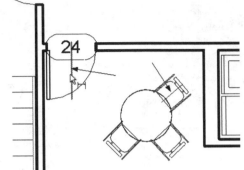 Select the **Angular** dimension tool from the ribbon.

14. Select the mid-line of the third Chair-Stacking and the mid-line of Door 24.

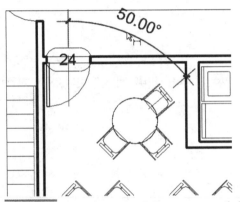

15. The angle should measure 50.00°.

16. 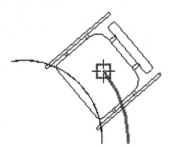 Select the **Measure** tool from the Quick Access toolbar.

17. Select the center of the chair as the start point.

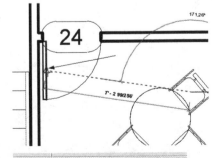

18. Select the midpoint of the interior side of Door 24.

19. | Total Length: | 7' - 2 99/256" | The distance is displayed below the ribbon.

20.

Views (all)
  Floor Plans
    Ground Floor
      Ground Floor - Record Storage
    Lower Roof

Open the **Ground Floor- Record Storage** floor plan view.

21.

Select the file cabinet located at the upper right of the room.

Record Storage

109

10

12

22. Select the **Array** tool from the ribbon.

23. Enable the **Linear** option.

24. ☑ Group and Associate   Number: 4

Enable **Group and Associate**. Set the Number to **4**.

25. Move To: ⦿ 2nd ○ Last   ☑ Constrain

Enable Move to **2ⁿᵈ**. Enable **Constrain**.

26.

Select the top left endpoint as the base point.

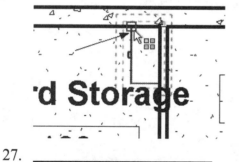

27.

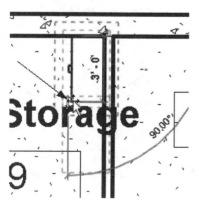

Select the bottom left endpoint as the distance between elements.

28.

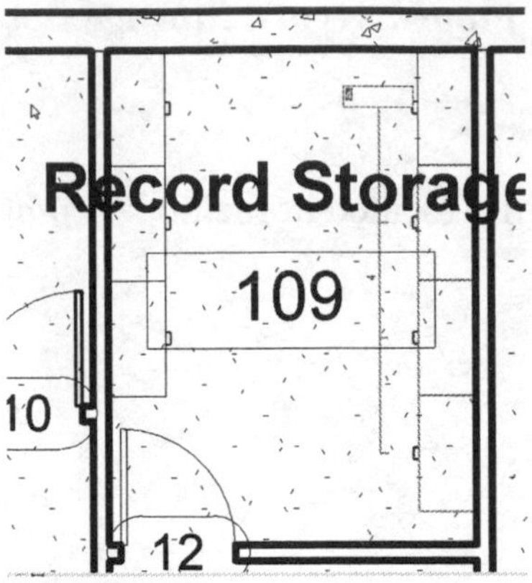

Record Storage

109

10

12

A preview will appear with the number 4.
Press **ENTER** to accept.
Left click anywhere in the display window to exit
the command.

29.

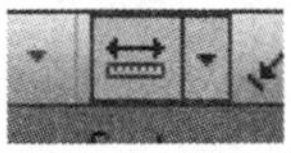

Select the **Measure** tool from the Quick Access toolbar.

30.

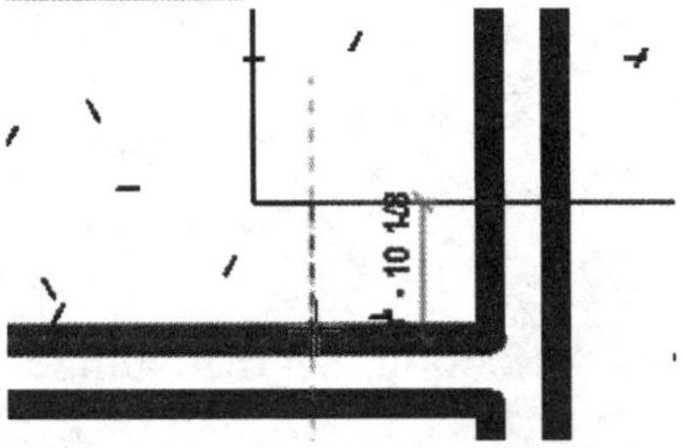

Measure from the bottom edge of the last file
cabinet to the interior side of the horizontal wall.

31.

Total Length:	0' - 10 1/8"

The value appears below the ribbon.

*This value can be copied and pasted into the
browser for the certification exam.*

32. Close without saving.

Exercise 2-3
Compound Walls

Drawing Name: **i_stacked_walls.rvt**
Estimated Time to Completion: 60 Minutes

Scope
Defining a compound wall structure

A compound wall has multiple vertical layers and/or regions. A layer is assigned to each row with a constant thickness and extends the height of the wall. A region is any shape in the wall that is situated in one or more layers. The region may have a constant or variable thickness.

Solution

1.   Open *i_stacked_walls.rvt*.

2.   Floor Plans   **Level 1**   Level 2   Site   Activate **Level 1**.

3.   Select the left vertical wall.

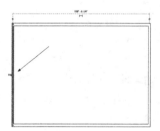

4.   Properties   Basic Wall   Exterior - Brick on CMU   Walls (1)   Edit Type   Select **Edit Type** on the Properties pane.

5.   Duplicate...   Select **Duplicate**.

6.   Name: Exterior - Concrete Foundation   Type **Exterior - Concrete Foundation**. Press **OK**.

7.

Type Parameters	
Parameter	**Value**
Construction	
Structure	Edit...
Wrapping at Inserts	Do not wrap

Select **Edit Structure**.

8.

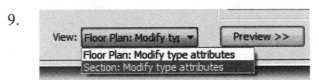

Expand the dialog by pressing the **Preview** button.

9.

View: Floor Plan: Modify typ ▼   Preview >>
Floor Plan: Modify type attributes
Section: Modify type attributes

Switch the view to **Section: Modify Type**.

10.

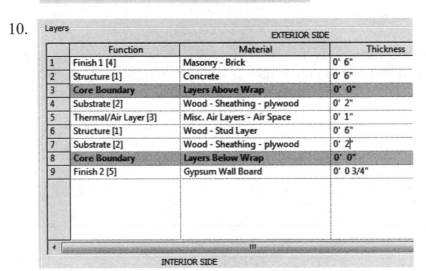

Layers

EXTERIOR SIDE

	Function	Material	Thickness
1	Finish 1 [4]	Masonry - Brick	0' 6"
2	Structure [1]	Concrete	0' 6"
3	**Core Boundary**	**Layers Above Wrap**	**0' 0"**
4	Substrate [2]	Wood - Sheathing - plywood	0' 2"
5	Thermal/Air Layer [3]	Misc. Air Layers - Air Space	0' 1"
6	Structure [1]	Wood - Stud Layer	0' 6"
7	Substrate [2]	Wood - Sheathing - plywood	0' 2"
8	**Core Boundary**	**Layers Below Wrap**	**0' 0"**
9	Finish 2 [5]	Gypsum Wall Board	0' 0 3/4"

INTERIOR SIDE

Add Layers as follows:
Layer 1: Finish 1 [4] Masonry - Brick 6″
Layer 2: Structure [1] Concrete 6″
Layer 3: Core Boundary
Layer 4: Substrate [2] Wood - Sheathing 2″
Layer 5: Thermal Air/Layer – Misc. Air Layers - Air Space 1″
Layer 6: Structure [1] Wood - Stud Layer 6″
Layer 7: Substrate [2] Wood - Sheathing 2″
Layer 8: Core Boundary
Layer 9: Finish 2 [5] Gypsum Wall Board 3/4″

11.

Split Region

Select **Split Region**.

12. 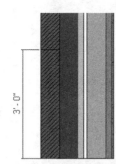   Cut the Layer 1: brick layer 3'-0" from the base.

Toggle the 2D button to see the hatch patterns on the layers.

13.

	Function	Material
1	Finish 1 [4]	Masonry - Brick
2	Structure [1]	Concrete

Highlight the **Layer 2: Concrete** Layer.

14. Assign Layers   Pick on the **Assign Layers** button.

15. 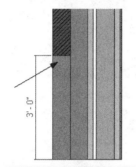   Select the lower region of the brick layer that was just split.

Left pick just below the cut line.

The upper region will now be brick and the lower region will be concrete.

It may take some practice before you are able to do this.

16. Split Region   Select **Split Region**.

17. 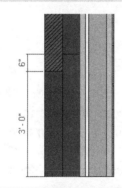   Cut the Layer 2: concrete 3'-6" from the base.

18.

EXTERIOR

	Function	Material
1	Finish 1 [4]	Masonry - Brick
2	Structure [1]	Concrete
3	**Core Boundary**	**Layers Above Wrap**
4	Substrate [2]	Wood - Sheathing - plywood

Highlight the **Layer 1: Masonry Brick** Layer.

19. Assign Layers   Pick on the **Assign Layers** button.

20. 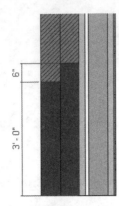 Highlight the Masonry brick layer.
Select the upper region of layer 2.

Left pick slightly above the cut line.

The upper region will now be brick and the lower region will be concrete.

It may take some practice before you are able to do this.

21.  Select **Modify**.

22. 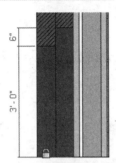 Select the base of the concrete Layer 1 component.

23.  A small lock will appear.
Click on the lock to unlock it.

24. Select the base of Layer 2: Concrete.
Click on the lock to unlock it.

25. Press **OK** to close the dialogs.

26.

Base is Attached	☐
Base Extension Distance	-3' 0"
Top Constraint	Unconnected
Unconnected Height	20' 0"
Top Offset	0' 0"

Select the wall with the Exterior - Concrete Foundation wall type.
In the Properties pane:
Set the Base Extension Distance to **-3' 0"**.
Left Click in the display window to release the selection.

27.  Set the display to Medium or Fine to see the wall layers.

28. 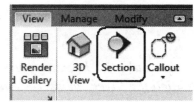 Activate the **View** ribbon.
Select the **Section** Tool from the Create panel.

29. 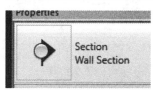 Set the section type to be a **Wall Section**.

30. 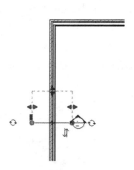 Place a small section on the wall you just defined.
Activate the section view.

31.  Set the display to Medium or Fine to see the wall layers.

32. 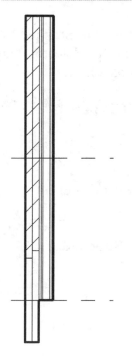 Note the concrete section is below the base level.
Select the wall.

33. You can use the grips to adjust the base depth of the concrete section.

This type of wall is called a compound wall, because you have split wall layers and modified the layers using regions.

34. Floor Plans
    Level 1
    Level 2
    Site

    Activate **Level 1**.

35. Stacked Wall
    Exterior - Brick Over CMU w Metal Stud
    Stacked Wall 1

    In the Project Browser, locate the two Stacked Wall types.

    These are the stacked walls available in the current project.

36. Select the south wall.

37.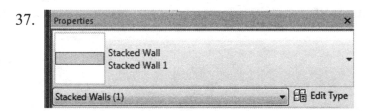

    Properties
    Stacked Wall
    Stacked Wall 1

    Stacked Walls (1)   Edit Type

    Switch the wall to **Stacked Wall 1** using the Type Selector on the Properties panel.

    Select **Edit Type**.

38. Duplicate...

    Select **Duplicate**.

39. Name: or - Brick with Concrete Foundation

    Rename **Exterior - Brick with Concrete Foundation**.
    Press **OK**.

40. Type Parameters

Parameter	Value
Construction	
Structure	Edit...

    Select **Edit** Structure.

41.

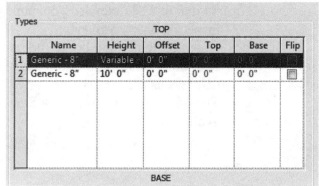

Note the stacked wall uses different layers going from Top to Base instead of Exterior to Interior.

Each layer is a wall type instead of a component material. These wall types are called subwalls.

42.

	Name	Height	Offset	
1	Exterior - Brick on Mtl. Stud	Variable	0' 0"	0'
2	Foundation - 36" Concrete	3' 0"	0' 0"	0'

Change Layer 1 to **Exterior Brick on Mtl. Stud**.
Change Layer 2 to **Foundation - 36" Concrete**.
Set the Height of Layer 2 to **3'-0"**.

43.

Insert

Select **Insert**. Position the new layer between the existing layers.

44.

	Name	Height	Offset	Top
1	Exterior - Brick on Mtl. Stud	Variable	0' 0"	0' 0"
2	Foundation - 12" Concrete	3' 6"	-0' 0 7/8"	0' 0"
3	Foundation - 36" Concrete	3' 0"	0' 0"	0' 0"

Set the new layer to:
Foundation- 12" Concrete.
Set the Height to **3' 6"**.
Set the Offset to **-7/8"**.

45.

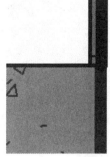

You can zoom into the preview window to check the offset value.

46.

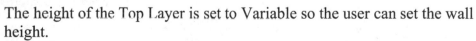

The height of the Top Layer is set to Variable so the user can set the wall height.
Press **OK** twice to exit the dialog.

Subwalls can be moved up or down the height of a stacked wall.

47.

Switch to a 3D view so you can inspect the two wall types.
Note that when you hover the mouse over the first wall you defined, it displays as a Basic Wall.
The other wall displays as a Stacked Wall.

48.

Activate the Level 1 view.
Select the North wall.

0' - 4"

49.

Basic Wall
Exterior - Siding with Wood St

Use the Type Selector drop-down to set the Type to **Exterior - Siding with Wood Stud**.

50.

Properties   ×

Stacked Wall
Exterior - Brick with Concrete Foundation

Stacked Walls (1)   Edit Type

Select the South Wall (the stacked wall).
Select **Edit Type**.

51.

Duplicate...

Select **Duplicate**.

52.

r - Siding with Concrete Foundation

Rename **Exterior - Siding with Concrete Foundation**. Press **OK**.

53.

Type Parameters

Parameter	Value
Construction	
Structure	Edit...

Select **Edit** Structure.

54.

	Name	Height	Offset	Top
1	Exterior - Siding with Wood Stud	Variable	0' 0"	0' 0"
2	Foundation - 12" Concrete	3' 6"	0' 1"	0' 0"
3	Foundation - 36" Concrete	3' 0"	0' 0"	0' 0"

Set Layer 1 to the new wall type: **Exterior - Siding with Wood Stud. Adjust the Offset for** Layer 2: Foundation -12" Concrete to **1"**. Press **OK** to close all dialogs.

55.  You can zoom into the preview window to check the offset. Press OK twice to close the dialogs.

56. 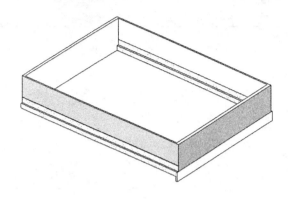 Switch to a 3D view.

    How many stacked walls are there?

    How many basic walls?

    What is the difference between a compound wall and a stacked wall?

    Is a compound wall a basic wall or a stacked wall?

57. Close without saving.

Exercise 2-4

Placing a Wall Sweep

Drawing Name: **walls.rvt**
Estimated Time to Completion: 20 Minutes

Scope
Placing a wall sweep.

Solution

1.  Activate **Level 1** Floor Plan.

2.  Select the **Wall** tool from the Architecture ribbon on the Build panel.

3. Set the wall type to **Exterior - Brick on Mtl. Stud** using the Type Selector on the Properties pane.

4. 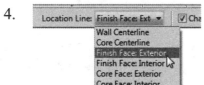 Set the Location Line to **Finish Face: Exterior**.

5. 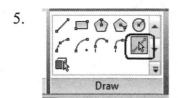 Select the **Pick Line** mode from the Draw panel.
   Select the four green lines.

   Note that when you pick the lines, the side of the line you use determines which side of the line is used for the exterior side of the wall.

6.

The lines should be aligned to the exterior side of the walls.

| 1/8" = 1'-0" | ⊠ | 🗇 ⟨ˣ 🔍ˣ 📐 📐 🔄 ♀ ◂ |

Set the Detail Level to **Medium**.

7.

Elevation

Activate the **View** ribbon.
Select the **Elevation** tool on the Create panel.

8.

Place an elevation in the center of the room.
Right click and select **Cancel** to exit the command.

Place a check mark on each box to create an elevation for each interior wall.

9.

Elevations (Building Elevation)
East
Elevation 1 - a
Elevation 1 - b
Elevation 1 - c
Elevation 1 - d
North
South
West

In the Project Browser, you will see that four elevation views have been created.

10.

Views : Elevation : Elevation 1 - a

If you hover your mouse over a triangle, a tooltip will appear with the name of the linked view.

11.

Elevations (Building Elevation)
East
East Interior
North
North Interior
South
South Interior
West
West Interior

Rename the elevation views to East Interior, North Interior, South Interior and West Interior.

*Pressing **F2** is a shortcut for Rename.*

12.

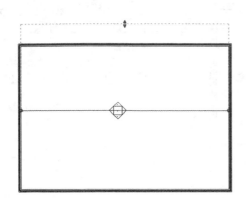

If you pick on the triangle part of the elevation, you will see the view depth (Far Clip Offset) of that elevation view.

13.

Elevations (Building Elevation)
- East
- East Interior
- North
- North Interior
- South
- **South Interior**
- West
- West Interior

Activate the **South Interior** View.

14. Use the grips to extend the elevation view beyond the walls.

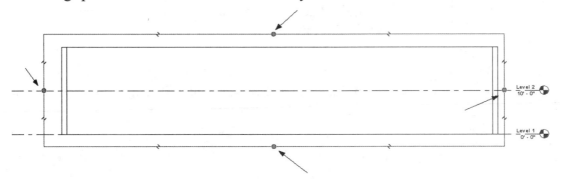

15.

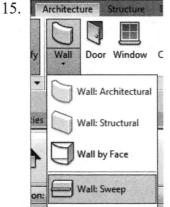

Activate the **Architecture** ribbon.
Select the **Wall Sweep** tool.

The Wall Sweep tool is only available in elevation, 3D or section views.

16. Place the sweep so it is toward the top of the wall.

17.  Select the wall sweep.
In the Properties pane, adjust the Offset from Level to **18′ 0″**.

18. 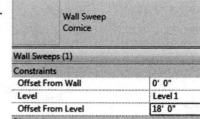 Switch to a 3D view.

19. Select the top corners of the view cube to orient the view so you can see the wall sweep.

20.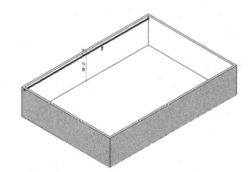

Select the wall sweep that was placed.
It will **highlight** when selected.

21. Select **Add/Remove Walls** from the ribbon.
Select the other walls.
Orbit around to inspect.

22. Save as *ex2-2.rvt*.

Exercise 2-5

Create a Wall Sweep Style

Drawing Name: **ex2-2.rvt**
Estimated Time to Completion: 15 Minutes

Scope
Creating a wall sweep style.
Loading a Profile.

Solution:

1. Elevations (Building Elevation)
    - East
    - East Interior
    - North
    - North Interior
    - South
    - **South Interior**
    - West
    - West Interior

    Activate the **South Interior** View.

2. Load Family   Load as Group
    Load from Library

    Activate the **Insert** ribbon.
    Select **Load Family**.

3. File name: "Crown 1" "Base 3"
    Files of type: All Supported Files (*.rfa, *.adsk)

    Load the following profiles:
    Base-3.rfa
    Crown 1.rfa

    These can be downloaded from this link:
    www.SDCpublications.com.

    You can load more than one file at a time by holding down the CTL key.

5. Press **Open**.

6. 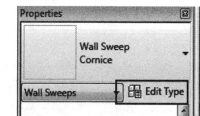 Activate the **Architecture** ribbon.
Select the **Wall Sweep** tool.

The Wall Sweep tool is only available in elevation, 3D or section views.

7. Select **Edit Type** from the Properties pane.

8. Select **Duplicate**.

9. Enter **Base Moulding** in the Name field.
Press **OK**.

10. Set the Profile to **Base 3 : 3 1/2″ x 9/16″**.
Press **OK** to exit the dialog.

11. Place the baseboard on the bottom of the wall.

12. Save as *ex2-3.rvt*.

Exercise 2-6

Create a Custom Profile

Drawing Name: **ex2-3.rvt**
Estimated Time to Completion: 20 Minutes

Scope
Creating a custom profile.
Using the custom profile in a wall sweep.

Solution

1. 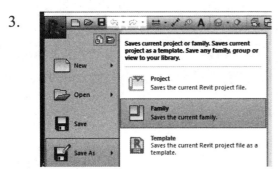 Browse under Profiles in the Project Browser.
Locate the **Base 3** profile family.

2.   Right click on the **Base 3** family.
Select **Edit**.

3.   Save the file as *Base 4.rfa*.

4.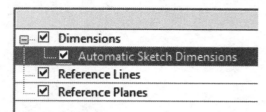

    Bring up the Visibility/Graphics dialog. *You can do this by typing VV.* Select the Annotation Categories tab. Enable all the Annotation Categories. Press **OK**.

5.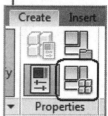

    Activate the Create ribbon. Select the **Family Types** tool on the Properties pane.

6.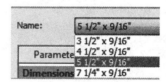

    Note that several sizes are available for this profile. Use the Apply button to see how the profile changes depending on the size selected. Set the size to **5 1/2" x 9/16"**. Press **OK** to close the Types dialog.

7.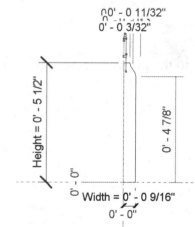

    Modify the profile. I eliminated the offset on the left and simplified the top. You can delete any unnecessary dimensions, but be sure to keep the Height and Width Dimensions. Verify that the profile still flexes properly using the different types.

8. Save the new profile.

9.

    Activate the Modify ribbon. Select **Load into Project**.

10. Close the family file.

11. 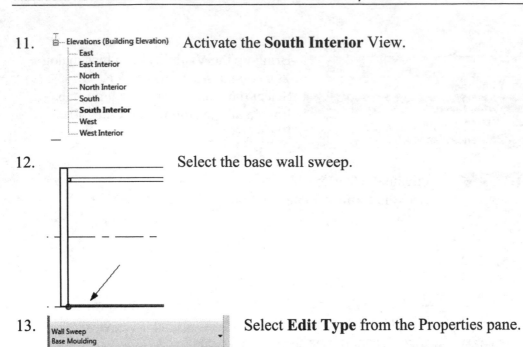 Activate the **South Interior** View.

Elevations (Building Elevation)
    East
    East Interior
    North
    North Interior
    South
    South Interior
    West
    West Interior

12. Select the base wall sweep.

13. Select **Edit Type** from the Properties pane.

Wall Sweep
Base Moulding

(1)   Edit Type

14. Select **Base 4: 5 1/2″ x 9/16″** for the Profile.

Default Setbac 0′ 0″

Construction

Profile   Base 4 : 5 1/2″ x 9/16″

This is the new profile you just created and loaded into the project. Press **OK**.

15. Save as *ex2-4.rvt*.

Exercise 2-7

Dividing a Wall into Parts

Drawing Name: **wall_parts.rvt**
Estimated Time to Completion: 45 Minutes

Scope
Use of parts to apply materials to a wall

Solution

1. Elevations (Building Elevation)
    - East
    - North
    - **South**
    - West

    Activate the **South** elevation view.

2.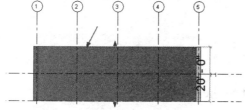

    Select the wall so it highlights.

3. Modify | Walls

    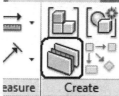

    ... Create

    Select the **Create Parts** tool on the Create panel.

4.

    Divide Parts

    Select **Divide Parts** on the Part panel.

5. Specify a new Work Plane
    - Name `<none>`
    - ⦿ Pick a plane
    - Pick a line and use the work plane it was sketched in

    Enable **Pick a plane**.
    Press **OK**.

6. 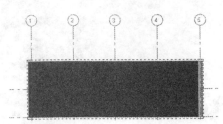   Pick the front of the wall for the work plane.

7. 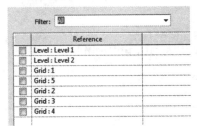   Select the **Add** tool from the Divided Parts panel.

8.    Select **Intersecting References** from the References panel.

9.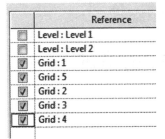

Set Filter to **All**.

This allows you to see Grids and Levels.

10.

Reference
Level : Level 1
Level : Level 2
Grid : 1
Grid : 5
Grid : 2
Grid : 3
Grid : 4

Place a check next to the **Grids**.

Press **OK**.

11.

Select the **Green Check** on the Mode panel to finish dividing the parts.

12. 3D Views   Switch to a 3D view.
         {3D}

13.

Scale Value 1:	90
Detail Level	Medium
Parts Visibility	Show Parts

In the Properties pane:
Set the Detail Level to **Medium**.
Set the Parts Visibility to **Show Parts**.

14.

Select the second panel/part.

15.

Original Type	Exterior - Brick on CMU
Material By Original	☐
Material	Concrete - Cast-in-Place Concrete
Construction	Finish

In the Properties panel:
Uncheck **Material by Original**.
Left click in the Material column to assign a material.

16.

Select the **Concrete - Precast Concrete** material.
Press **OK**.
Left click in the window to release the selection.

17.

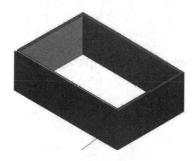

The wall changes to show the new material.

18.

Select the fourth panel/part.

19.

Original Type	Exterior - Brick on CMU
Material By Original	☐
Material	Concrete - Cast-in-Place Concrete
Construction	Finish

In the Properties panel:
Uncheck **Material by Original**.
Left click in the Material column to assign a material.

20.  Select the **Concrete - Precast Concrete** material.
Press **OK**.
Left click in the window to release the selection.

21.  The wall changes to show the new material.

22. Hold down the CTRL key and select the two concrete panels.

23. Select the **Divide Parts** tool from the Part panel.

24. 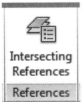 Select **Intersecting References** from the References panel.

25. 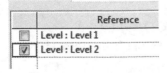 Place a check next to the **Level 2**.
Press **OK**.

26. Select the **Green Check** on the Mode panel to finish dividing the parts.

27. The concrete panels are now divided into two sections.

Shown in wireframe so you can see the divisions easily.

28.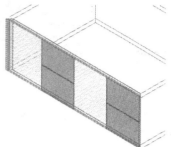

Hold down the CTRL key and select the two top sections of the concrete panels.

29. In the Properties panel:
Left click in the Material column to assign a material.

30.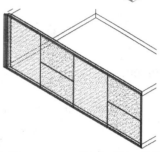

Select the **Masonry-Stone** material.
Press **OK**.
Left click in the window to release the selection.

31. 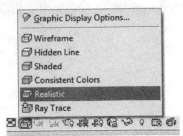   Set the display to **Realistic**.

32.    The new materials are assigned.

33. 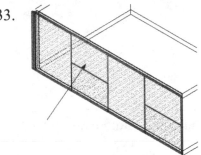   Select the top part indicated.

Note that no grips are available when a part is selected.

Shown in wireframe so you can see the divisions easily.
Use the TAB key to cycle through selection choices.

34. 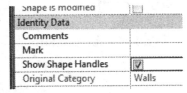   Enable **Show Shape Handles** on the Properties pane.

35. 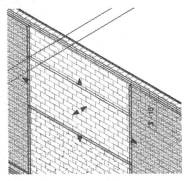   Use the top shape handle to lower the material 3' -10" from the top of the wall.
Left click in the window to release the selection.

36.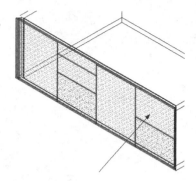

Select the top part indicated.

Note that no grips are available when a part is selected.

Shown in wireframe so you can see the divisions easily.

37.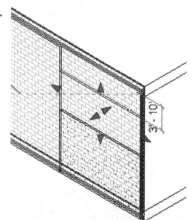

Enable **Show Shape Handles** on the Properties pane.

38.

Use the top shape handle to lower the material 3' -10" from the top of the wall.

39.

Switch to a **Realistic** display.
Orbit the model to inspect your new wall.

40.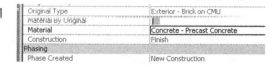

Select the bottom part of the second panel.

41.

Click in the Material column.

42.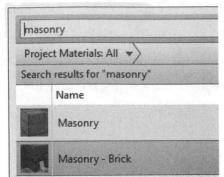

Set the Material to **Masonry -Brick**.
Press **OK**.

43.

Hold down the Control key and select the first panel, the bottom part of the second panel and the third panel.

44.

Merge Parts

Select **Merge Parts**.

Left click to release the selection.

45.

Select the far right lower part of the wall.

46.

Divide Parts

Select **Divide Parts**.

47.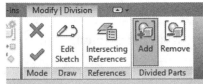

Enable **Add** on the Divided Parts panel.

48. 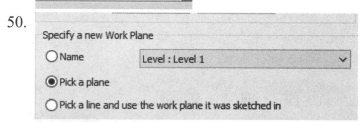 With Add enabled, select **Edit Sketch**.

49. 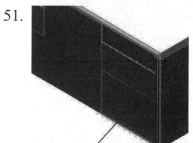 Select the **Set** tool on the Work Plane panel.

50. 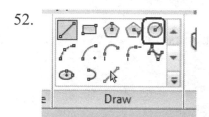 Enable **Pick a Plane**.
Press **OK**.

51. 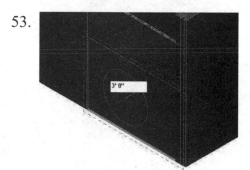 Select the front face of the wall part which is being edited.

52. Select the **Circle** tool.

53. Draw a 3'-0" radius circle.

54. 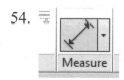 Use the dimension tool to place a horizontal and vertical dimension to locate the circle.

55. Select the circle.
This will allow you to modify the dimensions.
Change the horizontal dimension to 8' 6".
Change the vertical dimension to 4' 6".

56. 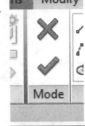 Select Green check to finish the sketch.

57. Select the circle that was placed.

58.

Original Type	Exterior - Brick on CMU
Material By Original	☐
Material	Concrete - Precast Con
Construction	Finish

Left click in the Material field on the Properties pane.

59. 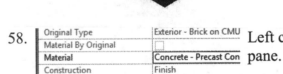 Assign a new material: **Masonry – Glass Block**.

60.  Inspect the wall.

61. Save as *ex2-6.rvt*.

Exercise 2-8
Creating an In-Place Mass

Drawing Name: **in_place_mass.rvt**
Estimated Time to Completion: 60 Minutes

Scope
Use of in-place masses to create a conceptual model

Solution

1. Floor Plans — Activate the **Site** plan view.
    Level 1
    Level 2
    Level 3
    Level 4
    Level 5
    Level 6
    Level 7
    Level 8
    Level 9
    Level 10
    Site

2.
    ture Insert Annotate Analyze **Massing & Site** Coll
    In-Place Place Curtain Roof Wall Floor Toposurface
    Mass Mass System
    al Mass   Model by Face

    Activate the **Massing & Site** ribbon.

    Select the **In-Place Mass** tool from the Conceptual Mass panel.

3. Revit has enabled the Show Mass mode, so the newly created mass will be visible.

    To temporarily show or hide masses, select the Massing & Site ribbon tab and then click the Show Mass button on the Massing panel.

    Masses will not print or export unless you make the Mass category permanently visible in the View Visibility/Graphics dialog.

    Revit displays a message indicating that visibility of masses has been turned on.

    Press **Close**.

4. Name: Building 1

    Name your first mass **Building 1**.

    Press **OK**.

5. 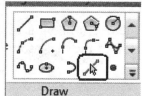   Select the **Pick Line** tool from the Draw panel.

6. 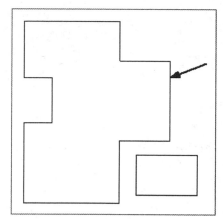   Pick the lines for the building in the upper right quadrant.

7. 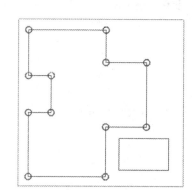   When you select the sketch, it should form a closed boundary.

Make sure there are no overlapping lines.

8. 3D Views   Switch to a 3D view.
       {3D}

9. 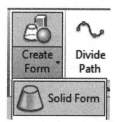   Select the sketch.

Select **Form→Create Form→Solid Form**.

10. Elevations (Building Elevation)   Switch to the **East** elevation.
       East
       North
       South
       West

2-41

11. There are two dimensions. The bottom dimension displays the overall height of the mass. The top dimension displays the distance from the top of the mass to the level above it.

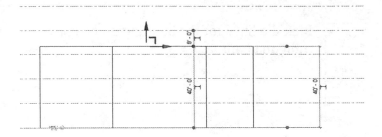

12.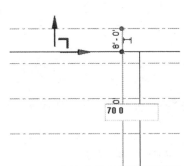

    Click on the bottom dimension controlling the overall height of the mass.
    Change it to **70′ 0″**.

    Press **ENTER**.
    Left click in the display window to release the selection.

13.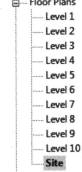

    Select **Finish Mass** from the In-Place Editor panel.

14. Floor Plans
        Level 1
        Level 2
        Level 3
        Level 4
        Level 5
        Level 6
        Level 7
        Level 8
        Level 9
        Level 10
        Site

    Activate the **Site** plan view.

15. Massing & Site

    Activate the **Massing & Site** ribbon.

16. In-Place Mass

    Select the **In-Place Mass** tool from the Conceptual Mass panel.

17.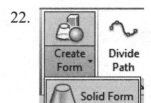

Name your mass **Building 2**.
Press **OK**.

18.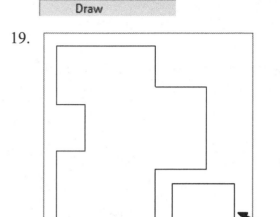

Select the **Pick Line** tool from the Draw panel.

19.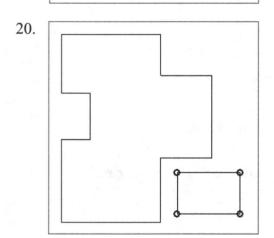

Pick the lines for the small building in the upper right quadrant.

You can also use the rectangle tool.

20.

When you select the sketch, it should form a closed boundary.

Make sure there are no overlapping lines.

21. Switch to a **3D** view.

22. Select the sketch.

Select **Form→Create Form→Solid Form**.

23.
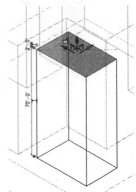
Select the Blue Z-axis.

Drag the building up until the dimension displays **70′-0″**.

Left click in the display window to release the selection.

24.

Select **Finish Mass** from the In-Place Editor panel.

25.
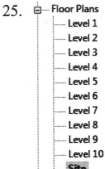
Activate the **Site** plan view.

26.

Activate the **Massing & Site** ribbon.

27.
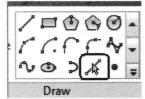
Select the **In-Place Mass** tool from the Conceptual Mass panel.

28.
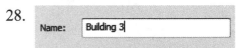
Name your mass **Building 3**.
Press **OK**.

29.
Select the **Pick Line** tool from the Draw panel.

30.

Pick the lines for the building in the upper left quadrant.

Select both the inner and outer boundaries to place lines.

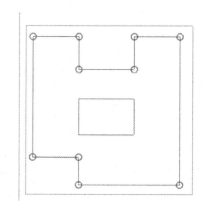

31.

When you select the sketch, it should form a closed boundary.

Make sure there are no overlapping lines.

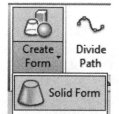

32. Switch to a **3D** view.

33. Select the outer sketch.

Select **Form→Create Form→Solid Form**.

34.

Select the Blue Z-axis.

Drag the building up until the dimension displays **82'-0"**.

Left click in the display window to release the selection.

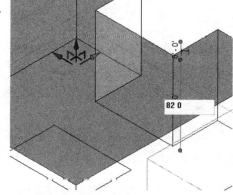

35. 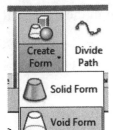 Select the inner sketch.

Select **Form→Create Form→Void Form**.

36. 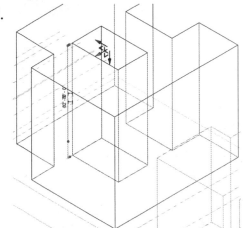 Select the Blue Z-axis.

Drag the void down until the dimension displays **82′-0″**.

Left click in the display window to release the selection.

37. Select **Finish Mass** from the In-Place Editor panel.

38. 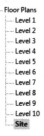 Activate the **Site** plan view.

39.  Activate the **Massing & Site** ribbon.

40. Select the **In-Place Mass** tool from the Conceptual Mass panel.

41. Name: Building 4   Name your mass **Building 4**.
Press **OK**.

42.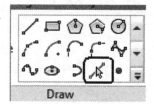

Select the **Pick Line** tool from the Draw panel.

43.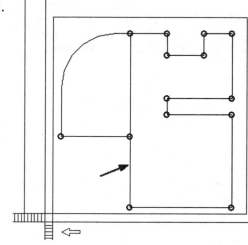

Pick the lines for the building in the lower left quadrant.

You will need to use the TRIM tool from the Modify panel to trim the lower left side of the sketch.

44.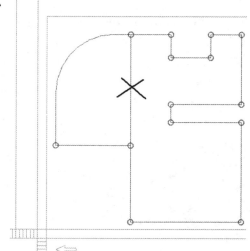

When you select the sketch, it should form a closed boundary.

Make sure there are no overlapping lines. Arrows indicate the sketch components. I have x'd out the line which needs to be trimmed out.

45. Switch to a **3D** view.

46. 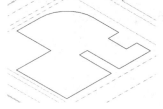 Select the sketch.

47. 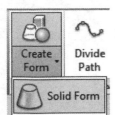 Select **Form→Create Form→Solid Form**.

48. 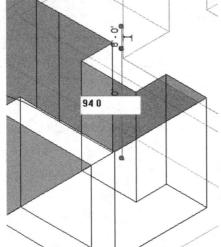 Select the Blue Z-axis.

Drag the building up until the dimension displays **94'-0"**.

Left click in the display window to release the selection.

49.  Select **Finish Mass** from the In-Place Editor panel.

50. 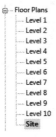 Activate the **Site** plan view.

51. Activate the **Massing & Site** ribbon.

52. Select the **In-Place Mass** tool from the Conceptual Mass panel.

In-Place
Mass

53.  Name the mass **Towers**.

Press **OK**.

54. 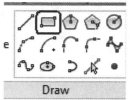 Select the **Rectangle** tool from the Draw panel.

55. 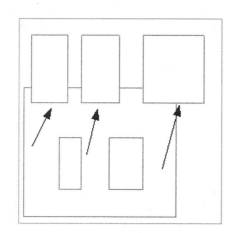 Trace over the top three rectangles in the fourth quadrant.

56.  Switch to a **3D** view.

57. 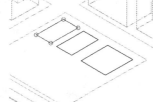 Select the left rectangle.

You can only use one closed polygon at a time for a solid form.

58. 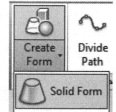 Select **Form→Create Form→Solid Form**.

59. 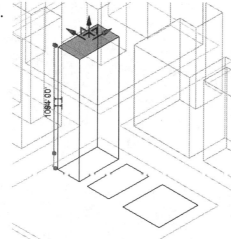 Select the Blue Z-axis.

Drag the building up until the dimension displays **106'-0"**.

Left click in the display window to release the selection.

60. 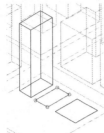 Select the middle rectangle.

61. 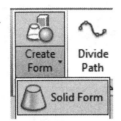 Select **Form→Create Form→Solid Form**.

62. 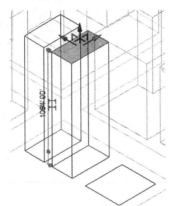 Select the Blue Z-axis.

Drag the building up until the dimension displays **106'-0"**.

Left click in the display window to release the selection.

63. 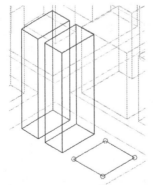 Select the right rectangle.

64. 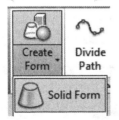 Select **Form→Create Form→Solid Form**.

65. 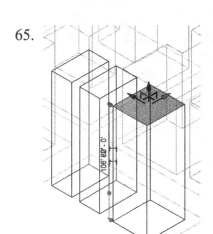 Select the Blue Z-axis.

Drag the building up until the dimension displays **106'-0"**.

Left click in the display window to release the selection.

66. Select **Finish Mass** from the In-Place Editor panel.

67. 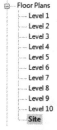 Activate the **Site** plan view.

68.  Activate the **Massing & Site** ribbon.

69. Select the **In-Place Mass** tool from the Conceptual Mass panel.

70. 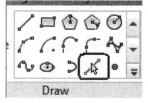 Name the mass **Building 5**.
Press **OK**.

71. Select the **Pick Line** tool from the Draw panel.

72.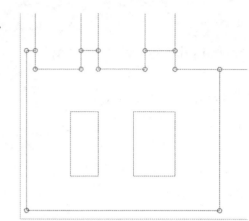

Pick the lines for the sketch in the lower right quadrant.
Use **Draw Line** to complete the sketch.

Disable Chain on the Options bar to make placing the lines easier.

Use the RECTANGLE tool to create two sketches for the internal rectangles.

These will be voids.

73.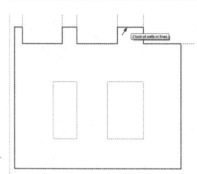

When you select one of the lines of the sketch, you should see the entire sketch highlight.

If you don't see a continuous loop, there are either missing lines or overlapping/duplicate lines.

Check the sketch for overlapping lines by deleting a line, then click UNDO if it is not a duplicate.

74. Switch to a **3D** view.

75.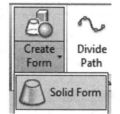

Select the outside boundary sketch.

Select **Form→Create Form→Solid Form**.

76.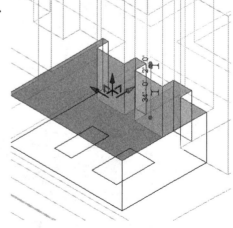

Select the Blue Z-axis.

Drag the building up until the dimension displays **34'-0"**.

Left click in the display window to release the selection.

77. 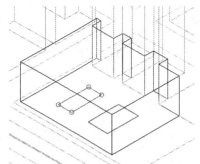 Select the left rectangle.

78. 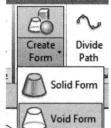 Select **Form→Create Form→Void Form**.

79. 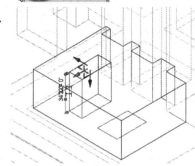 Select the Blue Z-axis.

Drag the building up until the dimension displays **34'-0"**.

Left click in the display window to release the selection.

80. 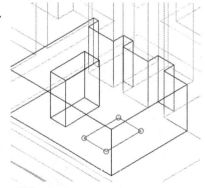 Select the right rectangle.

81. 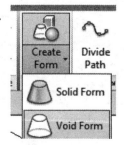 Select the sketch.
Select **Form→Create Form→Void Form**.

82.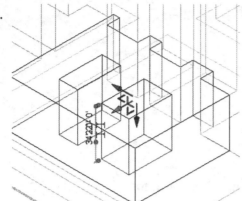

Select the Blue Z-axis.

Drag the building up until the dimension displays **34'-0"**.

Left click in the display window to release the selection.

83.

Select **Finish Mass** from the In-Place Editor panel.

84.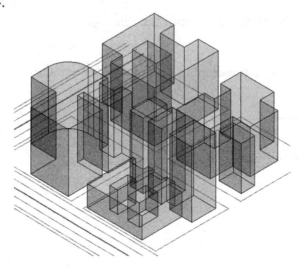

Close without saving.

Exercise 2-9

Editing an In-Place Mass

Drawing Name: **editing_masses.rvt**
Estimated Time to Completion: 30 Minutes

Scope
Editing in-place masses to develop a conceptual model

Solution

1.

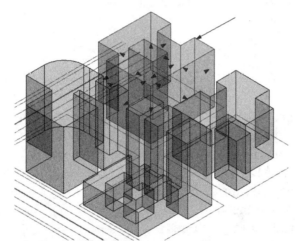

Select **Building 3** in the NW quadrant.

If you hover your mouse over a mass, it will display the mass name assigned.

2.  Select **Edit In-Place** from the Model panel.

3.

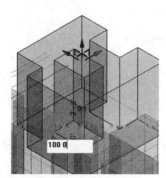

Use the TAB key to cycle through the selections until you have selected the top face.

Change the height of the mass to **100'-0"**.

4.

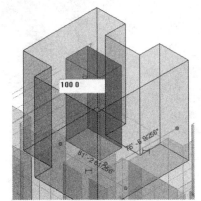

Use the TAB key to cycle through the selections until you have selected the void.

Change the height of the void to **100′-0″**.

5.

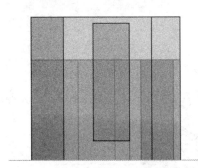

If you switch to a South elevation you can check the void to see if it really is aligned on the top and bottom.

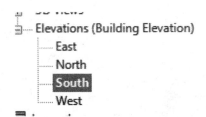

6.

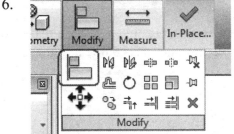

You can also use the ALIGN tool on the Modify panel to align the top of the void with the top of the solid form.
Select ALIGN. Use the TAB key to select the top of the solid form. Then, use the TAB key to select the top of the void.
Repeat for the bottom.

7.

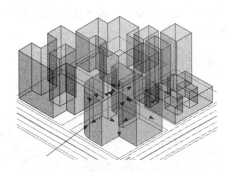

Select **Finish Mass** on the In-Place Editor panel when you are done editing the mass.

8.

Select Building 4.

9.  Select **Edit In-Place** from the Model panel.

10.  Rotate the view using the ViewCube.

11.  Select **Set Work Plane** from the Work Plane panel on the Create ribbon.

12. 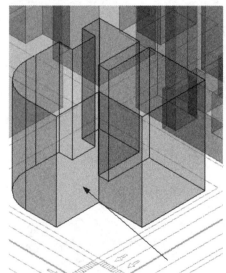 Select the face indicated.

13. 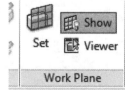 Select **Show Work Plane** from the Work Plane panel.

    The color of the selected face will change.

14. 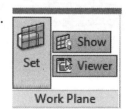 Select **Viewer** from the Work Plane panel.

15. 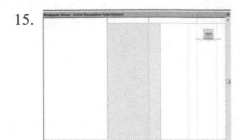 A window will open with a normal (perpendicular) view to the active work plane.

16. 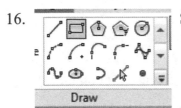 Select the **Rectangle** tool from the Draw panel.

17. 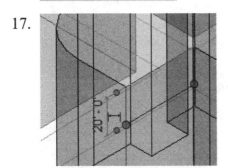 Draw a rectangle that is 20′ high using the Viewer window.

To adjust the dimension, select the bottom line and the temporary dimension will appear.

You can also use the ALIGN tool to set the sides of the rectangle collinear to the mass.

18. 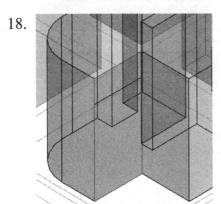 Check the placement of the rectangle in the 3D view.

Close the Viewer window.

19. 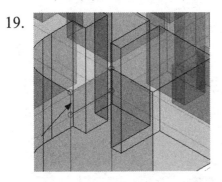 Select the sketch.

20. 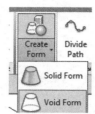 Select **Form→Create Form→Void Form**.

21. 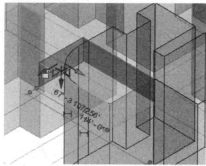 Use the Green Axis to drag the void form through the existing mass.

22.  Select **Finish Mass** from the In-Place Editor panel.

23. 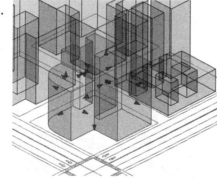 Select **Building 4**. (This is the building you just modified.)

24. Select **Mass Floors** from the Model panel.

25. 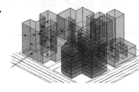 Enable all the Levels.

Press **OK**.

26. 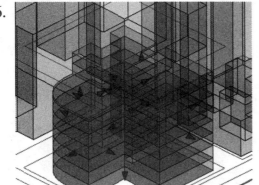 Floors are placed at each level.

27. Select Building 3.

28.  Select **Edit In-Place** from the Model panel.

29. Orient your view cube as shown.

30.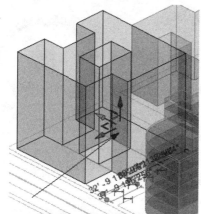

Select the face indicated.

You should see the shape handle axis tool.

You can use the TAB key to cycle through the selection until the face is highlighted then left click.

31.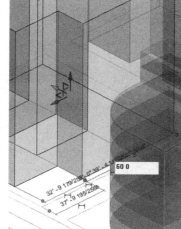

Adjust the dimension using the red axis.

It may be difficult to see the distance. You can also select the temporary dimension and set it to 60' 0". The arrows indicate the edge which is being adjusted.

32.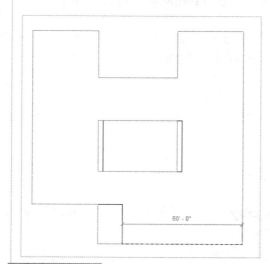

To verify that the face was moved properly, switch to the Site floor plan view.

Then use the MEASURE tool to verify the dimension.

33.

Select **Finish Mass** from the In-Place Editor panel.

34. Save the file as *ex2-8.rvt*.

Exercise 2-10

Mass Properties

Drawing Name: **new**
Estimated Time to Completion: 15 Minutes

Scope
Modifying a conceptual mass

Solution

1. Start a new project using the Architectural template.

2. 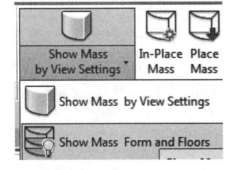 Activate the Massing & Site ribbon.

   Enable **Show Mass Form and Floors** from the Conceptual Mass panel.

3.  Activate the Architecture ribbon.

   Select the **Component→Place a Component** tool on the Build panel.

4.  Select **Load Family** from the Mode panel.

5.  Browse to the *Mass* folder.

6. Select the *Rectangle-Blended* mass.

   | File name: | Rectangle-Blended |
   | Files of type: | All Supported Files (*.rfa, *.adsl |

   Press **Open**.

7. 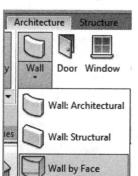 On the ribbon:

   Select **Place on Work Plane** on the Placement panel.

8. Place in the window.

   Right click and select Cancel twice to exit the command.

9. Switch to a 3D view.

10. 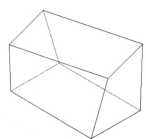 Activate the Architecture ribbon.

    Select the **Wall→Wall by Face** tool on the Build panel.

11. 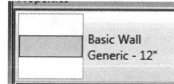 Set the wall type to **Generic - 12"** using the Type Selector.

12.

Select the face indicated.

Left click in the window to exit the command.

13.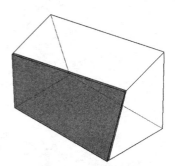

Select the wall you just placed.

What is the area of the wall?

14.

Dimensions	
Volume	1494.54 CF
Area	1503.14 SF

You should see the Volume and Area in the Properties pane.

15. Close without saving.

Things to remember about Masses:

- The default material for a mass is 5 percent transparent.
- Masses will not print unless the category is enabled in Visibility/Graphics Overrides.
- Masses are created from a single closed profile.
- Masses are a nested entity. In order to modify the profile, you have to open the mass up for editing and then open the desired form component up for editing.
- Masses can be comprised of multiple forms, a combination of voids and solids.

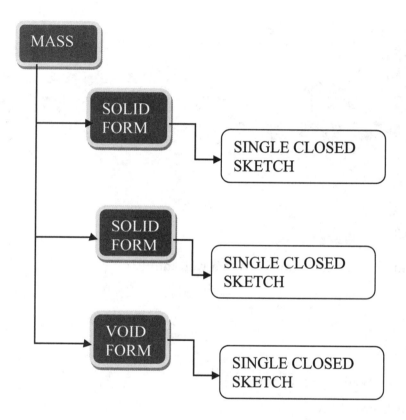

Certified User Practice Exam

1. Which of the following can NOT be defined prior to placing a wall?

   A. Unconnected Height
   B. Base Constraint
   C. Location Line
   D. Profile
   E. Top Offset

2. Identify the stacked wall.

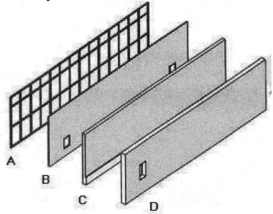

3. Walls are system families. Which name is NOT a wall family?

   A. BASIC
   B. STACKED
   C. CURTAIN
   D. COMPLICATED

4. Select the TWO that are wall type properties:

   A. COARSE FILL PATTERN
   B. LOCATION LINE
   C. TOP CONSTRAINT
   D. FUNCTION
   E. BASE CONSTRAINT

5. Select ONE item that is used when defining a compound wall:

   A. MATERIAL
   B. SWEEPS
   C. GRIDS
   D. LAYERS
   E. FILL PATTERN

6. Use this key to cycle through selections:

    A. TAB
    B. CTRL
    C. SHIFT
    D. ALT

7. The construction of a stacked wall is defined by different wall _____.

    A. Types
    B. Layers
    C. Regions
    D. Instances

8. To change the structure of a basic wall you must modify its:

    A. Type Parameters
    B. Instance Parameters
    C. Structural Usage
    D. Function

9. If a stacked wall is based on Level 1 but one of its subwalls is on Level 7, the base level for the subwall is Level _____.

    A. 7
    B. 1
    C. Unconnected
    D. Variable

Certified Professional Practice Exam

1. A Mass Face:

    A. Can be used to generate a building element, such as a wall or roof.
    B. Can be moved.
    C. Can be assigned a material.
    D. Can be deleted without deleting the mass.

2. When working with a mass, you can use levels to define mass _____

    A. Roofs
    B. Floors
    C. Ceilings
    D. Walls

3. To create a mass that is unique in a project, use the _____ Mass tool.

    A. In-Place
    B. Component
    C. System
    D. By Face

4. Selecting a work plane, when placing a mass:

    A. Automatically changes the relative coordinate system
    B. Changes the project location
    C. Changes the level
    D. Determines the depth of an extrusion

5. Select THREE element types that can be created using mass faces:

    A. Doors
    B. Walls
    C. Levels
    D. Floors
    E. Roofs

6. Select TWO methods used to create a conceptual design using mass families:

    A. Go to the Applications Menu and select New→Conceptual Mass.
    B. Go to the Massing & Site ribbon and select Place Mass.
    C. Go to the Applications Menu and select New→Family
    D. Go to the Massing & Site ribbon and select In-Place Mass
    E. Go to the Applications Menu and select New→Project

7. In order to place a wall or floor on a mass face, the face must be:

    A. Horizontal
    B. Vertical
    C. Either Horizontal or Vertical
    D. Curved or spherical
    E. None of the above

8. To divide a floor or wall into parts, you can use the following (select all that apply):

    A. Lines
    B. Levels
    C. Grids
    D. Circles
    E. Arcs

9. To display parts in a view:

    A. Go to the Massing & Site ribbon and select Show Mass.
    B. On the Properties pane: set Parts Visibility to Show Parts
    C. Go to the Visibility/Graphics dialog and enable Parts.
    D. Go to Temporary Hide/Isolate and Reset

10. To assign a different material to a part, select the part and:

    A. On the Properties pane: Enable Material by Original
    B. Right click and select Assign Material
    C. On the Modify ribbon, select Paint from the Geometry Panel.
    D. On the Properties pane: Uncheck Material by Original, then assign a material in the material field.

Answers:
    1) A; 2) B; 3) A; 4) D; 5) B, D, & E; 6) B & D; 7) A; 8) B & C; 9) B; 10) D

Component Families

This lesson addresses the following User and Professional exam questions:

- Load Component Families
- Adding Components to a Project
- Creating families

Users should be able to understand the difference between a hosted and non-hosted component. A hosted component is a component that must be placed or constrained to another element. For example, a door or window is hosted by a wall. You should be able to identify what components can be hosted by which elements. Walls are non-hosted. Whether or not a component is hosted is defined by the template used for creating the component. A wall, floor, ceiling or face can be a host.

Some components are level-based, such as furniture, site components, plumbing fixtures, casework, roofs and walls. When you insert a level-based component, it is constrained to that level and can only be moved within that infinite plane.

Components must be loaded into a project before they can be placed. Users can pre-load components into a template, so that they are available in every project.

Users should be familiar with how to use Element and Type Properties of components in order to locate and modify information.

There are three kinds of families in Revit Architecture:
- system families
- loadable families
- in-place families

System families are walls, ceilings, stairs, floors, etc. These are families that can only be created by using an existing family, duplicating, and redefining. These families are loaded into a project using a project template.

Loadable families are external files. These include doors, windows, furniture, and plants.

In-place families are components that are created inside of a project and are unique to that project.

Exercise 3-1
Level-Based Component

Drawing Name: **i_components.rvt**
Estimated Time to Completion: 10 Minutes

Scope
Moving a component from one level to the next.

Solution

1.
Floor Plans
    Ground Floor
    Lower Roof
    Main Floor
    Main Roof
    Site
    T. O. Footing
    T. O. Parapet

Activate the **Main Floor** floor plan.

2.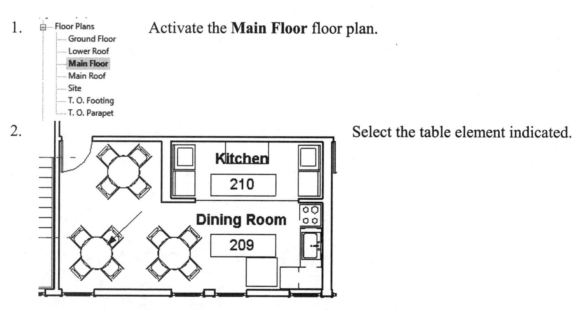

Select the table element indicated.

3.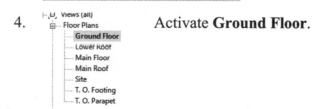

On the Properties pane:
Change the Level parameter to **Ground Floor**.

4.
Views (all)
    Floor Plans
        Ground Floor
        Lower Roof
        Main Floor
        Main Roof
        Site
        T. O. Footing
        T. O. Parapet

Activate **Ground Floor**.

5.

The table is in the same location on the Ground Floor.

6. Close the file without saving.

Exercise 3-2
Indentifying a Family

Drawing Name: **i_firestation_elem.rvt**
Estimated Time to Completion: 5 Minutes

Scope
Identify different elements and their families.

Solution

1. Open *i_firestation_elem.rvt*.

2. 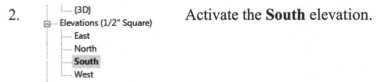 Activate the **South** elevation.

3. 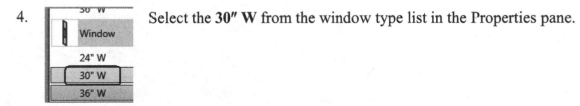 Select the second window from the left.

4. Select the **30″ W** from the window type list in the Properties pane.

    24″ W
    30″ W
    36″ W

5. Use the Measure tool from the Quick Access toolbar to check the distance between the outside edges of the two left windows.

6.

Check to see if you got the same value.

7.   Close without saving.

Exercise 3-3

Creating a Family

Drawing Name: **park bench.dwg**
Estimated Time to Completion: 80 Minutes

Scope
Create a Revit family from an AutoCAD symbol.

Solution

1. Close any open files.

2.  Go to **New→Family**.

3. Select the *Furniture.rft* from the *English_I* templates. Press **Open**.

    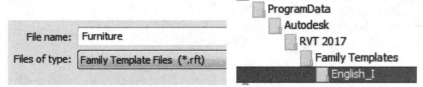

4. Activate the **View** ribbon.

    Select **Tile** under the Window panel.

5. Family1 - Elevation: Right   Double click the title for the **Elevation:Right** window to enlarge it.

6.

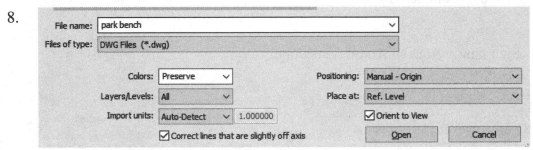

Activate the **Create** ribbon.

Select **Forms→Extrusion**.

7.

Activate the **Insert** ribbon.

Select the **Import CAD** tool from the Import panel.

8.

File name:	park bench	⌄
Files of type:	DWG Files (*.dwg)	⌄

Colors:	Preserve ⌄		Positioning:	Manual - Origin ⌄
Layers/Levels:	All ⌄		Place at:	Ref. Level ⌄
Import units:	Auto-Detect ⌄ 1.000000			☑ Orient to View
	☑ Correct lines that are slightly off axis		Open	Cancel

Locate the *park bench.dwg* file. Set the positioning to Manual-Origin. Press Open.

9.

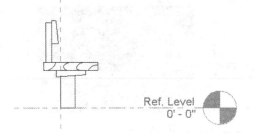

The park bench will appear as shown.

*Type **F** to set the view to extents to see the park bench.*

Ref. Level
0' - 0"

10.

Constraints	
Extrusion End	4' 0"
Extrusion Start	0' 0"
Work Plane	Reference Plane : Center Left/Right

On the Properties pane:

Set the Extrusion End to **4'-0"**.

11.

Materials and Finishes	
Material	<By Category>

Left click in the **Material** column to assign a material.

12.

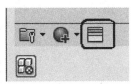

Launch the **Asset Browser**.

13. 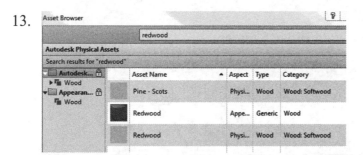 Type **redwood** in the search field.

14. Highlight the Default material in the Project Materials list. Highlight the Redwood material in the Asset Browser. Click on the far right of the Asset Browser to copy the asset browser properties over to the Default material definition.
    Close the Asset Browser.

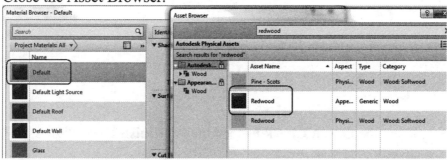

15. 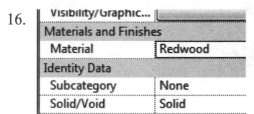 Rename the Default material **Redwood.**
    The Redwood material is displayed as a Document Material.
    Highlight the material and press **OK**.

16. 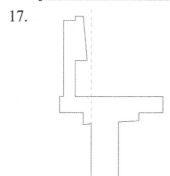 Redwood is listed in the Properties pane.

17. Clean up the sketch so that there are no self-intersecting lines.
    Use the DELETE, TRIM and SPLIT tools.

18. Select the **SPLIT** tool. Enable **Delete Inner Segment.**
    Select the two points where you want the segment removed.

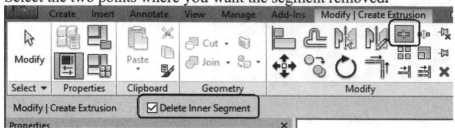

19. 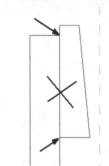 The arrows indicate the points to be selected and the x'd out section indicates the line segment to be deleted.

20. 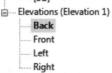 Select the **Green Check** under Mode to **Finish Extrusion.**

21. Switch to a 3D view. Switch to a Realistic display.

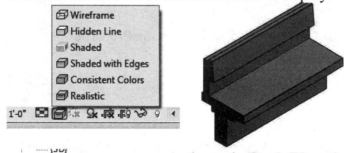

22. Activate the **Back Elevation** view.

23. Activate the **Create** ribbon.
    Select the **Set** tool under the Work Plane panel.

24. 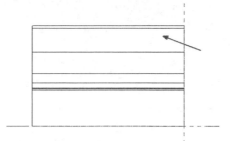 Enable **Pick a plane**.
Press **OK**.

25.  Select the front face of the back support.

26. 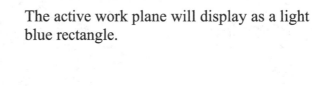 Select the **Show** tool on the Work Plane panel.

27. 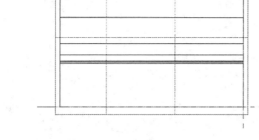 The active work plane will display as a light blue rectangle.

28. 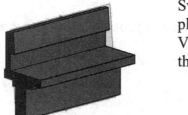 Switch to a 3D view and turn on the active work plane.
Verify that the active work plane is the front face of the back support.

29. Select **Forms→Extrusion**.

30. Select **Material** from the Properties Pane.

31 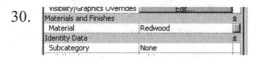 Select **Create New Material** at the bottom of the Material Editor dialog.

32.   Highlight the new material.
Right click and select **Rename**.

33.   Rename **Aluminum**.

34.   Launch the **Asset Browser**.

35.   Type **aluminum** in the search field.

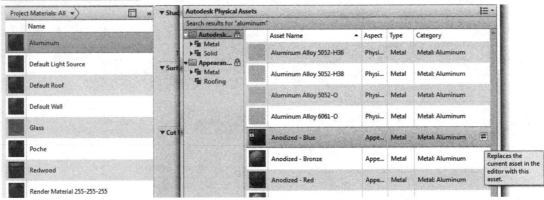

36.   Highlight the Aluminum material in the Material Editor.
Highlight the **Aluminum, Anodized Blue** material and **replace the current asset in the editor with this asset**.

37.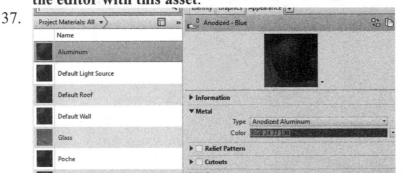

The aluminum material has been copied to the Project Materials list. Close the Asset Browser.

38.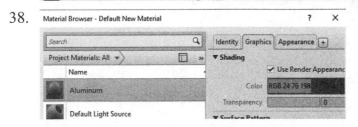

In the Material browser, select the Graphics tab.
Enable **Use Render Appearance**.

39. 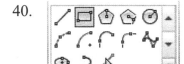 Highlight the Aluminum material to select.
Press **OK**. You should see the aluminum material listed in the Material field in the Properties pane.

 Set the Extrusion End to **1/4″**.

40. 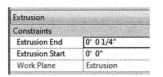  Return to a **Back** view.
Select the **Rectangle** tool from the Draw panel.

41.  Draw a rectangle on the seat face. Use dimensions to center it on the surface.

Remember temporary dimension values control the values of permanent dimensions.

42. Use the Fillet Arc tool to fillet the corners of the rectangle.

43. 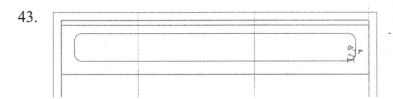 Set the corners to 1″ radius on the Options Bar.

44. Select the **Green Check** under Mode to **Finish Extrusion**.

45.    Switch to a 3D view to inspect your model.

46. 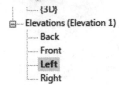   Activate the **Left Elevation** view.

47.   Activate the **Create** ribbon.
Select the **Reference Plane** tool.

48. 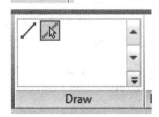   Select the **Pick** tool.

49. 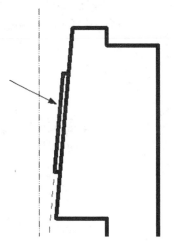   Pick the outside line of the aluminum plate.

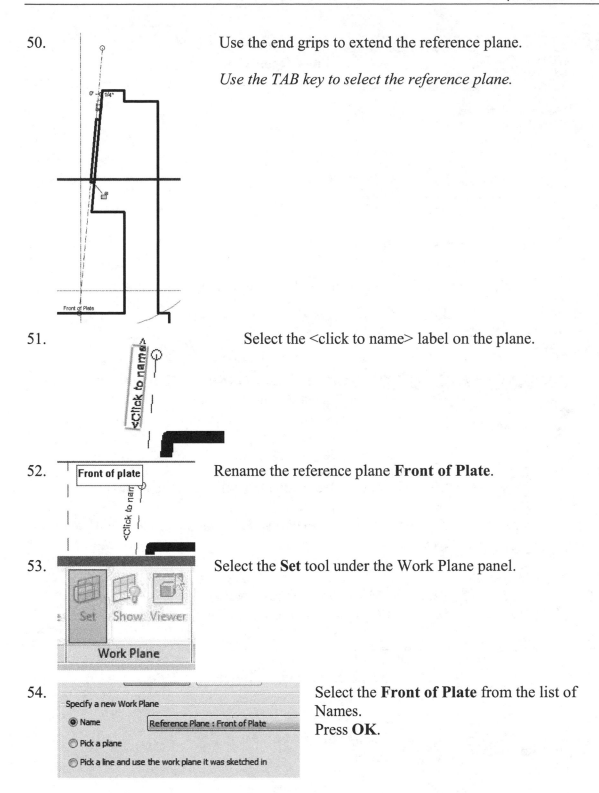

50.   Use the end grips to extend the reference plane.

Use the TAB key to select the reference plane.

51.   Select the <click to name> label on the plane.

52.   Rename the reference plane **Front of Plate**.

53.   Select the **Set** tool under the Work Plane panel.

54.   Select the **Front of Plate** from the list of Names.
Press **OK**.

55.

3D View: View 1
3D View: {3D}
Elevation: Back
Elevation: Front
Floor Plan: Ref. Level
Reflected Ceiling Plan: Ref. Level

Open View

Select **Elevation: Back**.
Press **Open View**.

56.
: Model
  Text

Select **Model Text** from the Model panel on the Create ribbon.

57.

Edit Text

In Memory of Our Fallen Soldiers

Type **In Memory of Our Fallen Soldiers** in the
Edit Text dialog.
Press **OK**.

58.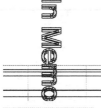

Click to place in the view.
Right click and select **Cancel** to exit the command.

The space bar is not available to rotate the text in the Model Text command.

59.

▼ 🔲 Edit Type

Select the Model text you just created.
Select **Edit Type** in the Properties Pane.

60. Duplicate...

Select **Duplicate**.

61.

Name: Model Text 2

Name the new type **Model Text 2**.
Press **OK**.

62.

Parameter	
Text	
Text Font	Arial
Text Size	0' 2"
Bold	☐
Italic	☐
Identity Data	

Change the Text Size to **2"**.
Press **OK**.

63.

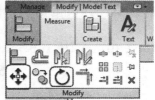

Use the Rotate and Move tools from the Modify panel to position the text.

64.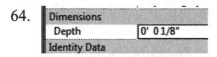

Set the Depth to **1/8″**.

65.

Select **Material** from the Properties Pane.

66.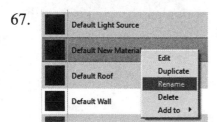

Select **Create New Material** at the bottom of the Material Editor dialog.

67.

Highlight the new material.
Right click and select **Rename**.

68.

Rename **Brass**.

69.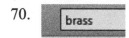

Launch the **Asset Browser**.

70.

brass

Type **brass** in the search field.

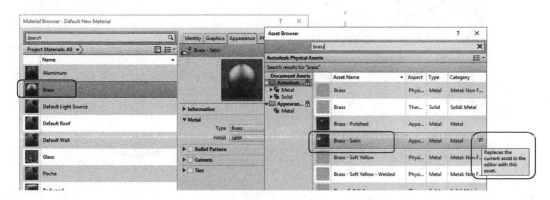

71. Highlight the **Brass** material in the Material Editor.
Highlight the **Brass – Satin** material in the Asset Browser and **replace the current asset in the editor with this asset**. Close the Asset Browser.

72. 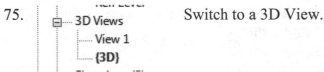 Select the Graphics tab in the Material Editor.
Enable **Use Render Appearance**.

73. The material has been copied to the Project Materials list. Highlight the brass material to select. Press **OK**.

74. 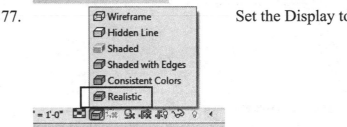 The material is listed in the Properties Pane.

75. Switch to a 3D View.

76. Turn off the visibility of the work plane by clicking on the Show Work Plane button.

77. Set the Display to **Realistic**.

78. The model so far.

79.  Activate the **Create** ribbon.
Select the **Set** tool under the Work Plane panel.

80. 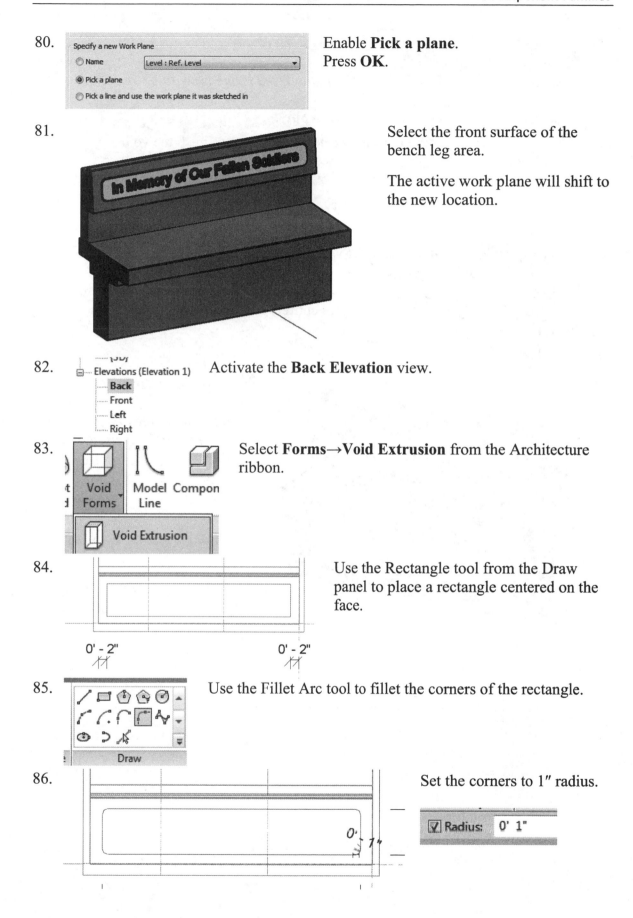 Enable **Pick a plane**.
Press **OK**.

81. Select the front surface of the bench leg area.

The active work plane will shift to the new location.

82. Activate the **Back Elevation** view.

83. Select **Forms→Void Extrusion** from the Architecture ribbon.

84. Use the Rectangle tool from the Draw panel to place a rectangle centered on the face.

0' - 2"   0' - 2"

85. Use the Fillet Arc tool to fillet the corners of the rectangle.

86. Set the corners to 1" radius.

Radius: 0' 1"

87.

Constraints	
Extrusion End	0' 8"
Extrusion Start	0' 0"
Work Plane	Extrusion
Identity Data	

Set the depth of the void extrusion **8"**.

Verify the Work Plane is set to Extrusion.

Press **Apply**.

88.

Select the **Green Check** under Mode to finish the void.

89.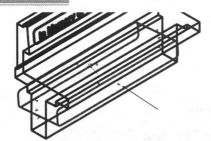

Switch to a wire frame view to see the void.

Use the grips to adjust the void position so it goes entirely through the park bench.

90.

Save as *park bench.rfa*.

This is a loadable family which can be inserted and used in any project.

Creating a Door Family

I assign this exercise as a homework assignment in my classes.

Scope

1. Use the Generic Model-Face Based template for the door panel.

2. Create the materials for the frame/door panel/glazing.

Add the following materials to the family document:

Mahogany
Canvas Paint (R244,G236,B215)
Frosted Glass

3. Create material parameters for the frame/door panel/glazing.
    Link the material parameters with the materials created.

4. Create the door panel - one extrusion for the exterior side and one extrusion for the interior side.

5. Assign the materials to the extrusions.

6. Open the Door template to start a new door family.

Create three door sizes as family types:

6' 10" x 3' 10"
6' 10" x 3' 2"
7' 2" x 3' 10"

7. Place a reference line to host the door panel.

8. Insert the door panel family.

9. Flex the model.

10. Add symbolic lines for the door swing.

11. Test the model family.

Exercise 3-4
Creating a Door Panel

Drawing Name: **new**
Estimated Time to Completion: 1 hours 30 minutes

Scope
Create a door panel using the Generic Model template
Create extrusions
Create voids
Define materials
Assign materials to extrusions
Define custom parameters

Solution

1. 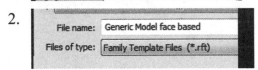 In order to use a door swing, the easiest way is to create the door panel as a generic model face based and insert it into a door family.

    Go to **New →Family**.

2. 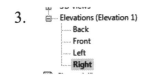 Select **Generic Model face based**.
    By selecting face based, we can control the position of the door using a reference line.

3. Elevations (Elevation 1)
    ┊---- Back
    ┊---- Front
    ┊---- Left
    ┊---- **Right**

    Activate the **Right** Elevation.

4. 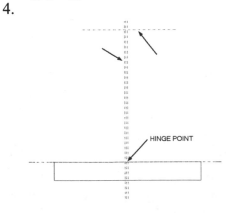 Add a horizontal reference plane and two vertical reference planes to the **left** of the existing reference plane.

    The intersection of the pre-existing work planes control the insertion point of the family when it is placed. We want to insert the door panel at the hinge point of the door.

HINGE POINT

5.

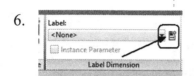

Add an EQ dimension between the vertical reference planes.

Add an overall vertical dimension and an overall horizontal dimension.

6.

Select the horizontal overall dimension and then select **Create Label** from the ribbon.

7.

Type **Thickness** for Name.
Enable **Type**.
Press **OK**.

8.

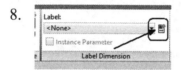

Select the vertical overall dimension and then select **Create Label** from the ribbon.

9.

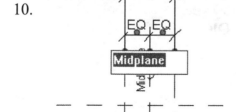

Type **Height** for Name.
Enable **Type**.
Press **OK**.

10.

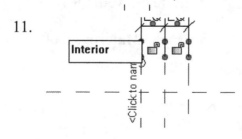

Name the middle vertical reference plane **Midplane**.

11.

Name the far left vertical reference plane **Interior**.

12. Activate the **Front** Elevation.

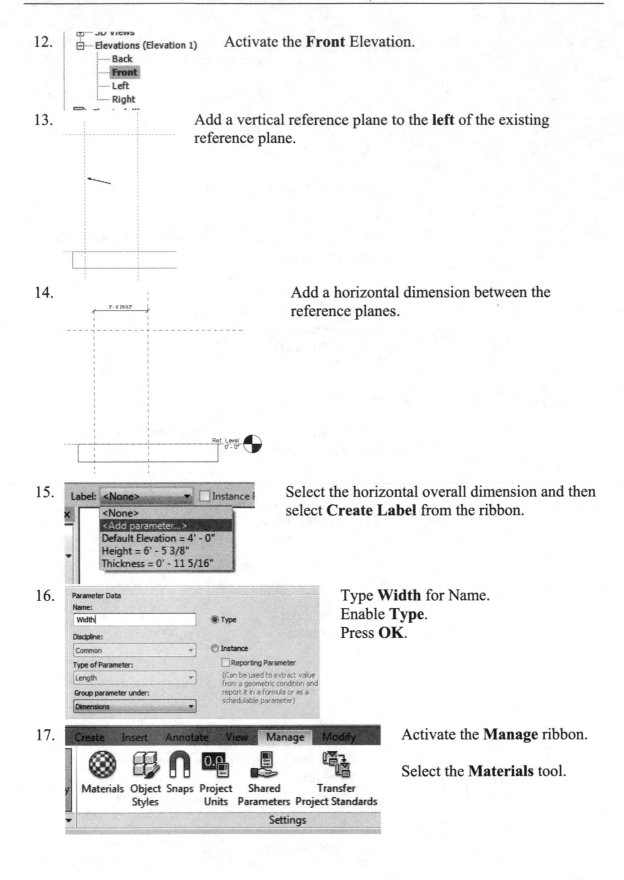

13. Add a vertical reference plane to the **left** of the existing reference plane.

14. Add a horizontal dimension between the reference planes.

15. Select the horizontal overall dimension and then select **Create Label** from the ribbon.

16. Type **Width** for Name.
Enable **Type**.
Press **OK**.

17. Activate the **Manage** ribbon.

Select the **Materials** tool.

Door Panel Exterior	Mahogany	These are the materials we want to create.
Door Panel Interior	Canvas Paint (R244,G236,B215)	
Glazing	Frosted	

18. 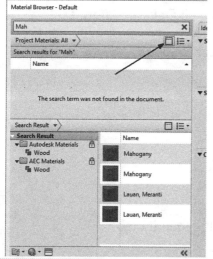 Enable **show library panel** if it is not visible.

19. 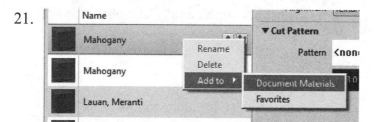 Do a search for Mahogany.

20. 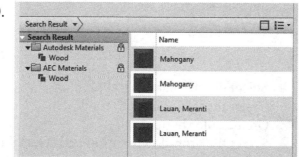 The two Mahogany materials in the Autodesk library will be displayed.

21. Highlight the first **Mahogany** material.

Right click and select **Add to→Document Materials**.

22. Select the **Graphics** tab.
Enable **Use Render Appearance**.

23.
Clear the search field.

Highlight the **Default** material.

Right click and select **Duplicate**.

24.
Highlight the duplicated Default material.

Right click and select **Rename**.

25.
Rename **Paint - Canvas**.

26.
Select the Identity tab.

Add a description.
Change the Class to Paint/Coating.
Add keywords to help if you need to search for this material.

27.
Select the Appearance tab.
Under the Generic section:
On the color button:
Select the drop-down button and select **Edit Color**.

28. Select one of the empty custom color slots.

29.
Change the RGB values to 244, 236, and 215.

Press **Add**.

30.

The custom color updates.

31.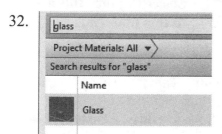

Select the **Graphics** tab.
Enable **Use Render Appearance**.

32.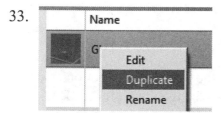

Do a search for Glass.

There is a Glass material in the active document.

33.

Name

Highlight **Glass**.
Right click and select **Duplicate**.

| Edit |
| Duplicate |
| Rename |

34.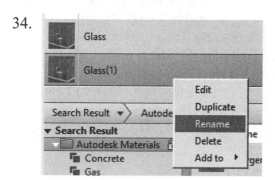

Highlight the copied glass material.

Right click and select **Rename**.

35. 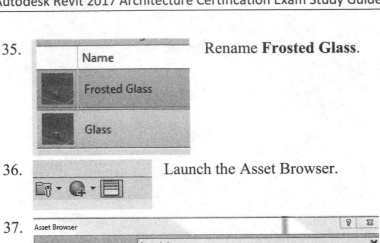 Rename **Frosted Glass**.

36. Launch the Asset Browser.

37. 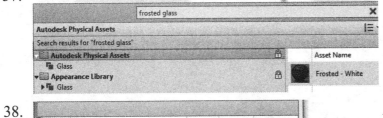 Perform a search for frosted glass in the Asset Browser.

38. Highlight the Glass Appearance row.
Select **Replace this asset in this material**.

39. Close the Asset Browser.

40. Select the Identity Tab:
In the Information section:
Change the Name to **Glass, Frosted - Blue**.

41. Select the Appearance tab.

In the Solid Glass Section:
    Change the Color to **Blue**.
In the Relief Pattern Section:
    Enable **Relief Pattern**.
Set the Type to **Rippled**.
In the Tint Section:
    Enable **Tint**.
Change the Tint Color to **Cyan**.
Press **OK**.

42.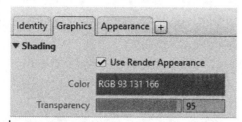

Select the **Graphics** tab.
Enable **Use Render Appearance**.

Close the dialog.

43.

Next we create parameters to control the material settings.

Activate the **Create** ribbon.

Select **Family Types**.

44.

Door Panel Exterior	Mahogany
Door Panel Interior	Canvas Paint (R244,G236,B215)
Glazing	Frosted

We need to create parameters for these materials.

45.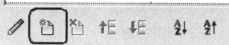

Select **New Parameter.**

46.

Parameter Data

Name:

Panel Exterior

Discipline:

Common

Type of Parameter:

Material

Group parameter under:

Materials and Finishes

○ Type

○ Instance

☐ Reporting Parameter

(Can be used to extract value from a geometric condition and report it in a formula or as a schedulable parameter)

For Name, type **Panel Exterior**.
For Type of Parameter, select **Material**.
Group parameter under: **Materials and Finishes**.
Enable **Type**.

Press **OK**.

By default, the Type of Parameter is set to Length. If you go too fast, you will not set the correct parameter type and you will have to delete your parameter and do it over.

47.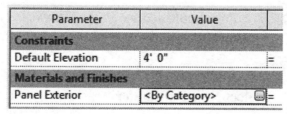

Under Materials and Finishes:
Locate the **Panel Exterior** parameter.

Click in the column to assign the material.

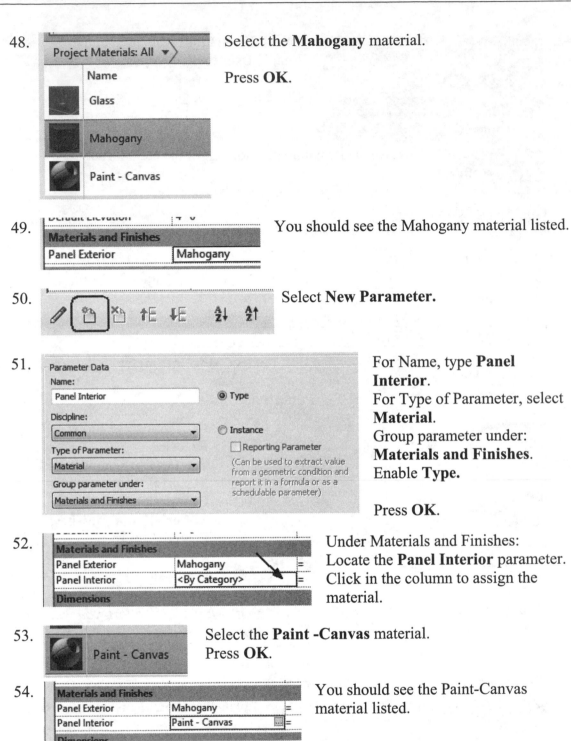

48. Select the **Mahogany** material.

Press **OK**.

49. You should see the Mahogany material listed.

50. Select **New Parameter.**

51. For Name, type **Panel Interior**.
For Type of Parameter, select **Material**.
Group parameter under: **Materials and Finishes**.
Enable **Type.**

Press **OK**.

52. Under Materials and Finishes:
Locate the **Panel Interior** parameter.
Click in the column to assign the material.

53. Select the **Paint -Canvas** material.
Press **OK**.

54. You should see the Paint-Canvas material listed.

55. Select **New Parameter.**

56.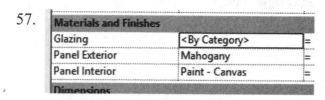

For Name, type **Glazing**.
For Type of Parameter, select **Material.**
Group parameter under: **Materials and Finishes**.
Enable **Type**.

Press **OK**.

57.

Materials and Finishes		
Glazing	<By Category>	=
Panel Exterior	Mahogany	=
Panel Interior	Paint - Canvas	=
Dimensions		

Under Materials and Finishes:
Locate the **Glazing** parameter.
Click in the column to assign the material.

58.

Name
Glass
Glass, Frosted - Blue
Mahogany

Select the **Glass, Frosted - Blue** material.

Press **OK**.

59.

Materials and Finishes	
Glazing	Glass, Frosted - Blue
Panel Exterior	Mahogany
Panel Interior	Paint - Canvas

You should see the Frosted Glass material listed.

Press **OK**.

60.
```
Elevations (Elevation 1)
    Back-Interior
    Front-Exterior
    Left
    Right
```

Rename the Front Elevation **Front-Exterior**.
Rename the Back Elevation **Back-Interior.**

I use both names in case a student doesn't rename their elevations, so I can troubleshoot their model.

61. Switch to the **Front-Exterior** Elevation view.
The exterior door panel will be extruded from the center reference plane to the exterior reference plane.

62.

Set   Show   Viewer

Work Plane

Activate the Create ribbon.
Select the **Set Work Plane** tool.

63.   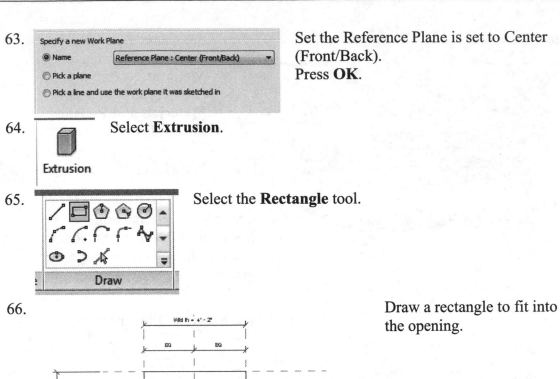   Set the Reference Plane is set to Center (Front/Back).
Press **OK**.

64.   Select **Extrusion**.

65.   Select the **Rectangle** tool.

66.   Draw a rectangle to fit into the opening.

67.

Width = 3' - 0 25/32"

6' 5 49/128

Enable all the locks.
This will ensure that the door panel size is controlled by the reference planes.

If you forget to lock the geometry, you can use the ALIGN tool to align the lines with the reference planes and lock each set.

68.

Materials and Finishes	
Material	<By Category>
Identity Data	

Select the small button on the far right of the Material parameter.

69.

Existing family parameters of compatible type:

<none>
Glazing
Panel Exterior
Panel Interior

Select the **Panel Exterior** material.

Press **OK**.

70.

Extrusion	▼	🔲 Edit Type
Constraints		
Extrusion End	1' 0"	
Extrusion Start	0' 0"	
Work Plane	Reference Plane : Center (Front/...	

Click the far right button on the Extrusion End parameter.

71.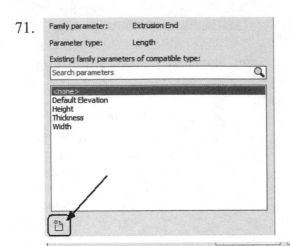

    Select Add parameter.

72.

    Parameter Data

    Name:

    Exterior Panel Thickness   ⦿ Type

    Discipline:

    Common ▾   ○ Instance

    Type of Parameter: ☐ Reporting Parameter

    Length ▾   (Can be used to extract value from a geometric condition and report it in a formula or as a schedulable parameter)

    Group parameter under:

    Dimensions ▾

    Enter **Exterior Panel Thickness**.

    Enable **Type**.

    Press **OK**.

73.

    Existing family parameters of con

    <none>
    Default Elevation
    Exterior Panel Thickness
    Height
    Thickness
    Width

    Select the **Exterior Panel Thickness** parameter.

    Press **OK**.

74.

    Properties

    Launch the **Family Types** dialog.

75. Locate the Exterior Panel Thickness parameter.
    In the Formula column: Type **Thickness/2**.
    Set the Thickness to 2". This will set the Exterior Panel Thickness to **1"**. Press
    OK.

Panel Interior	Paint - Canvas	=
Dimensions		
Exterior Panel Thickness	0' 1"	= Thickness / 2
Height	7' 0 53/64"	=
Thickness	0' 2"	=
Width	3' 3"	=
Identity Data		

76.

Constraints	
Extrusion End	0' 1"
Extrusion Start	0' 0"
Work Plane	Reference

In the Properties pane:
Note that the Extrusion End is set to 1".
It is grayed out because the value is controlled by a parameter.

77.

Green Check to finish the extrusion.

78.
- 3D Views
  - **View 1**
- Elevations (Elevati
  - Exterior

Switch to View 1.
Set the Visual Style to **Realistic**.
Verify that the material displays correctly.

79.
- Elevations (Elevation 1)
  - **Back-Interior**
  - Front-Exterior
  - Left
  - Right

Switch to an **Interior** Elevation view.

80.

Set   Show   Viewer

Work Plane

Activate the Create ribbon.

Select the **Set Work Plane** tool.

81.

Specify a new Work Plane
- Name   Reference Plane : Interior
- Pick a plane
- Pick a line and use the work plane it was sketched in

Set the Reference Plane to **Interior**.

Press **OK**.

82.

Extrusion

Select **Extrusion**.

83.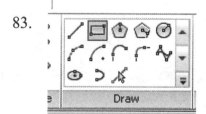

Draw

Select the **Rectangle** tool.

84.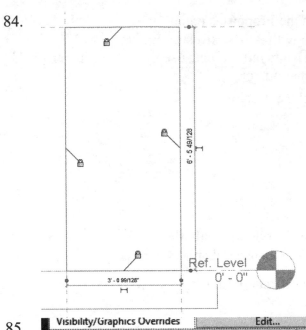

Draw a rectangle to fit into the opening.

Enable all the locks to ensure that the extrusion is controlled by the reference planes.

85.

Visibility/Graphics Overrides	Edit...
Materials and Finishes	
Material	<By Category>
Identity Data	

Select the small button on the far right of the Material parameter.

86.

Existing family parameters of compatible type:

<none>
Glazing
Panel Exterior
Panel Interior

Select the **Panel Interior** parameter.
This links the interior door panel to the material assigned to the Door Panel Exterior. Press **OK**.

87.

Extrusion	▾	Edit Type
Constraints	
Extrusion End	1' 0"
Extrusion Start	0' 0"
Work Plane	Reference Plane : Midplane

Click the far right button on the Extrusion End parameter.

88.

Family parameter: Extrusion End
Parameter type: Length

Existing family parameters of compatible type:

Search parameters

<none>
Default Elevation
Height
Thickness
Width

Select Add parameter.

89. 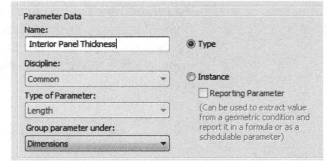 Enter **Interior Panel Thickness**.

Enable **Type**.

Press **OK**.

90. 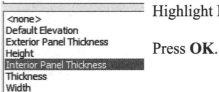 Highlight **Interior Panel Thickness**.

Press **OK**.

91.  Launch the **Family Types** dialog.

92. Locate the Interior Panel Thickness parameter.
In the Formula column: Type **Thickness/2**.
This will set the Interior Panel Thickness to **1"**.
Press **OK**.

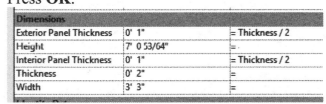

Dimensions		
Exterior Panel Thickness	0' 1"	= Thickness / 2
Height	7' 0 53/64"	=
Interior Panel Thickness	0' 1"	= Thickness / 2
Thickness	0' 2"	=
Width	3' 3"	=

93. 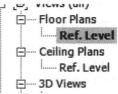 **Green Check** to finish the extrusion.

94. 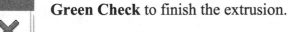 Switch to the **Ref. Level**.

95.

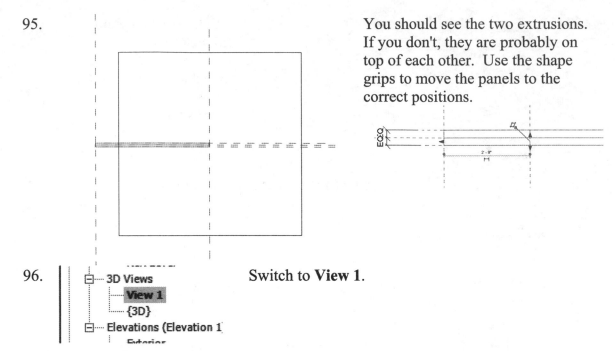

You should see the two extrusions. If you don't, they are probably on top of each other. Use the shape grips to move the panels to the correct positions.

96.

3D Views
    View 1
    {3D}
Elevations (Elevation 1
    Exterior

Switch to **View 1**.

97.

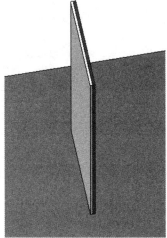

Verify that you see the two different materials.
Set the display to **Realistic** and orbit around the model.

98.

3D Views
Elevations (Elevation 1)
    Back-Interior
    Front-Exterior
    Left
    Right

Switch to an exterior elevation view.

99. Add reference planes to define locations for the glazing.
Remember to only dimension between reference planes. Do not select the extrusions!
Use EQ dimensions to center the middle horizontal reference plane and middle vertical reference planes.
The lower horizontal reference plane is 1'-0" above the Ref Level to define a kick plate area.

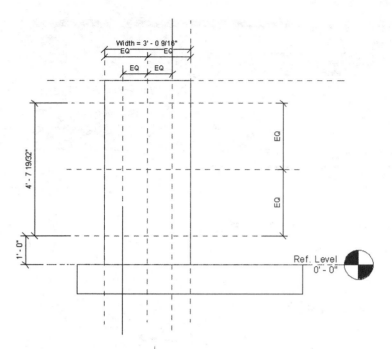

100.

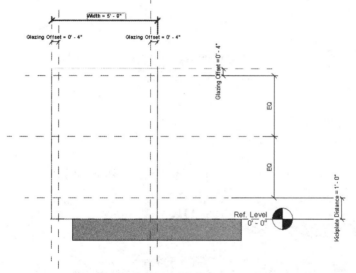

Add dimensions to offset the glazing 4" from the outside and the center.

Remember that the temporary dimensions set the values for the permanent dimensions. To change the values of the permanent dimensions, select the reference plane and then modify the temporary dimension.

Do not lock the dimensions or they won't flex properly.

I created a parameter called Glazing Offset and used it to apply the 4" dimension.
I created a parameter called Kickplate Distance and set it to 1'-0".
I then selected the dimensions and used the drop-down list to apply the parameters.

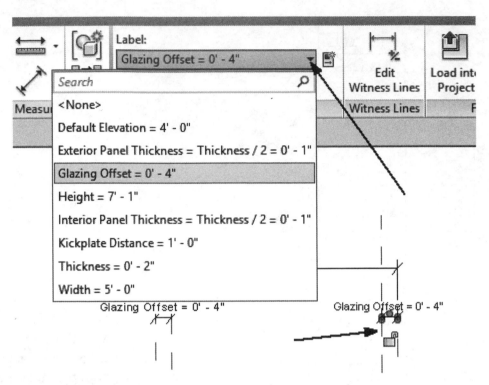

101.

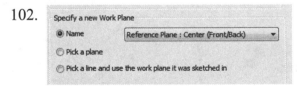

Activate the **Create** ribbon.
Select the **Set Work Plane** tool.

102.

Specify a new Work Plane

⦿ Name   Reference Plane : Center (Front/Back) ▾

◯ Pick a plane

◯ Pick a line and use the work plane it was sketched in

Set the Reference Plane to Center (Front/Back).
Press **OK**.

103.

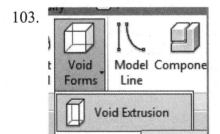

Select **Void Forms→Void Extrusion**.

First we create an opening for the glass, then we place the glass.

104.

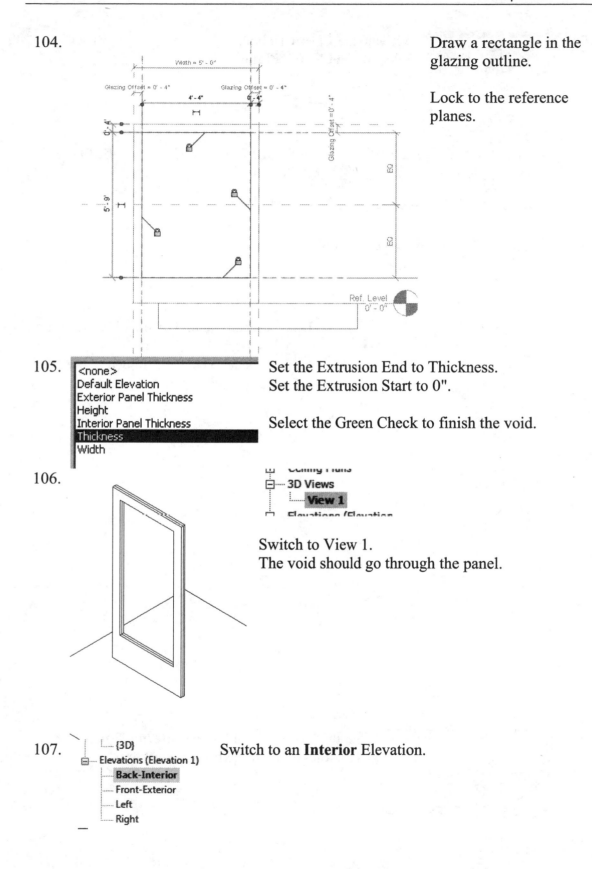

Draw a rectangle in the glazing outline.

Lock to the reference planes.

105.

<none>
Default Elevation
Exterior Panel Thickness
Height
Interior Panel Thickness
Thickness
Width

Set the Extrusion End to Thickness.
Set the Extrusion Start to 0".

Select the Green Check to finish the void.

106.

Switch to View 1.
The void should go through the panel.

107.

Switch to an **Interior** Elevation.

108.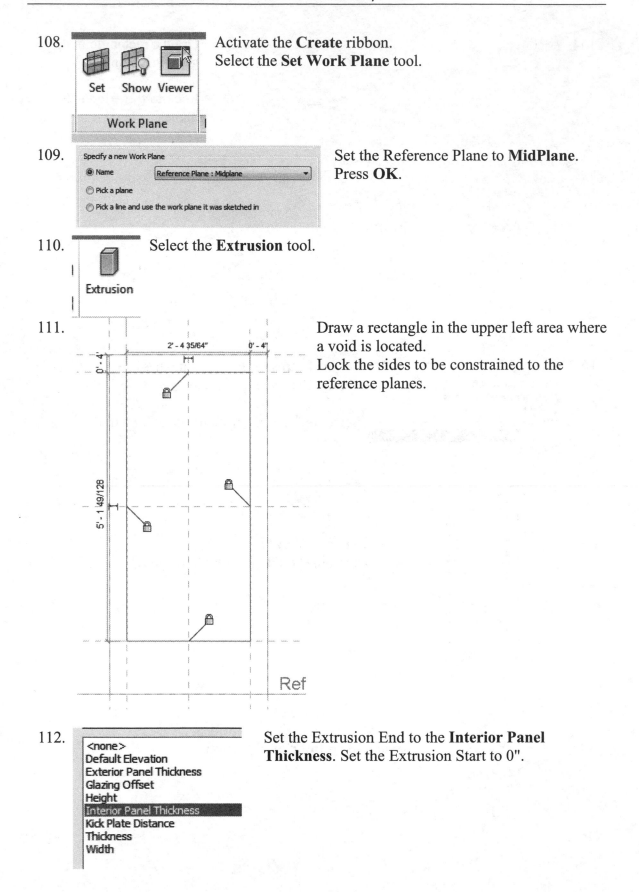
Activate the **Create** ribbon.
Select the **Set Work Plane** tool.

109.
Set the Reference Plane to **MidPlane**.
Press **OK**.

110.
Select the **Extrusion** tool.

111.
Draw a rectangle in the upper left area where a void is located.
Lock the sides to be constrained to the reference planes.

112.
Set the Extrusion End to the **Interior Panel Thickness**. Set the Extrusion Start to 0".

113. Set the Material to the **Glazing** parameter.

114. Select Green Check to finish the extrusion.

115. 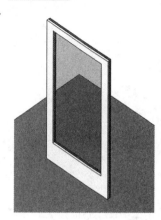 Switch to View 1 to inspect your door.
Save as **Door Panel Mahogany**.rfa.

Next we create the door family and insert the door panel.

Exercise 3-5

Create a Nested Door Family

Drawing Name: **new**
Estimated Time to Completion: 1 hours 15 minutes

Scope
Create a door family
Insert a generic model into a door family
Apply constraints to control the generic model's position and size
Link the generic model parameters to the door family parameters

Solution

1. 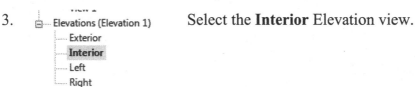 In order to use a door swing, the easiest way is to create the door panel as a generic model face based and insert it into a door family.
When you insert a family into another family, it is considered a *nested family*.
Go to **New** →**Family**.

2. 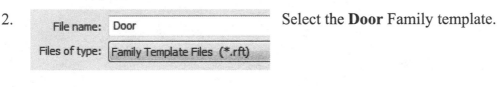 Select the **Door** Family template.

3. Elevations (Elevation 1)
   Exterior
   Interior
   Left
   Right

   Select the **Interior** Elevation view.

4.

Width = 3' - 0"

EQ

EQ

Height = 7' - 0"

Ref. Level

The template already has some parameters set up.

5.

3D Views

View 1

Activate View 1.

6.

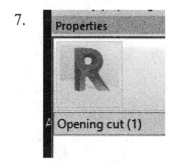

Select the rectangle.

7.

Properties

R

Opening cut (1)

If you look at the Properties pane, you see this is the opening for the door, not the door panel.

8. 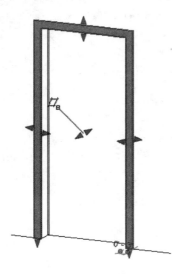 Select the extrusion surrounding the door.

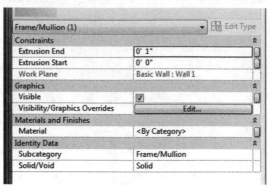

In the Properties pane, it is designated as Frame/Mullion.

9. Switch the view to the other side.

Note that there is a frame/mullion already created for both sides.

10. Select the **Family Types** tool.

11.  Under Family Types: Select **New Type**.

Family Types is where we define the different sizes of the door.

12. Enter the Door size. Press **OK**.

Name: Exterior - 6' 10" x 3' 10"

13. | | |
| --- | --- |
| **Dimensions** | |
| Height | 6' 10" |
| Width | 3' 10" |
| Rough Width | |
| Rough Height | |
| Thickness | |

Enter the correct Height and Width.

Press **Apply**.

You should see the door opening flex in the display window.

14.

Under Family Types: Select **New Type**.

Family Types is where we define the different sizes of the door.

15.

Name: Exterior - 6' 10" x 3' 2"

Enter the Door size.
Press **OK**.

16. | | |
| --- | --- |
| **Dimensions** | |
| Height | 6' 10" |
| Width | 3' 2" |
| Rough Width | |

Enter the correct Height and Width.
Press **Apply**.
You should see the door opening flex in the display window.

17.

Under Family Types:
Select **New Type**.
Family Types is where we define the different sizes of the door.

18.

Name: Exterior - 7' 2" x 3' 10"

Enter the Door size.
Press **OK**.

19. | | |
| --- | --- |
| **Dimensions** | |
| Height | 7' 2" |
| Width | 3' 10" |
| Rough Width | |

Enter the correct Height and Width.
Press **Apply**.

You should see the door opening flex in the display window.

20. Press **OK** to close the Family Types dialog.

21. Floor Plans
    Ref. Level

Activate the **Ref Level** view.

22.

Reference Line   Reference Plane

Datum

Activate the **Create** ribbon.
Select the **Reference Line** tool on the Datum panel.

The reference line can be used to host inserted elements.

23. 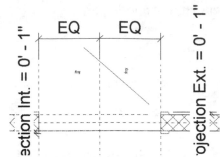   Draw an angled line above the door opening.

24. 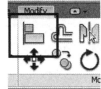   We need to align the ENDPOINT of the reference line with the horizontal plane and the left vertical plane.

Select the **ALIGN** tool.

25. 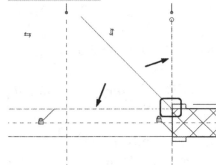   Select the horizontal plane then select the ENDPOINT of the reference line.

Use the TAB key if needed to cycle through the selections.

26. 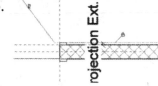   **Do not lock the alignment.**

27. 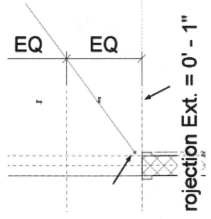   Select the Right reference plane and the ENDPOINT of the reference line to align.

Use the TAB key if needed to cycle through the selections.

You may need to move the reference line into position before setting the alignment lock.

28. Lock the alignment.

29. 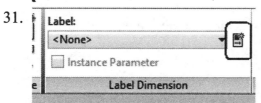 Select the **Angular Dimension** tool.

30. 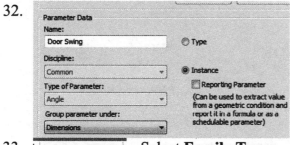 Place an angular dimension between the reference line and the exterior plane.

31. Select the angular dimension and select **Add parameter** from the Ribbon

Label:

<None>

☐ Instance Parameter

Label Dimension

32. Type **Door Swing** for the parameter name. Enable **Instance**.

Parameter Data

Name:

Door Swing

Discipline:

Common

Type of Parameter:

Angle

Group parameter under:

Dimensions

○ Type

◉ Instance

☐ Reporting Parameter

(Can be used to extract value from a geometric condition and report it in a formula or as a schedulable parameter)

Users can control the door swing value if it is an instance value.

Press **OK**.

33. Select **Family Types**.

Properties

34.

Construction Type	
Dimensions	
Height	7' 2"
Width	3' 10"
Door Swing (default)	90.000°
Rough Width	

Change the Door Swing to **90 degrees**. Press **Apply**.

The reference line should flex.

Try a few different angle values to verify that the reference line shifts properly.

If the reference line does not flex properly, delete it and try to lock it into place again.

Press **OK** to close the dialog.

35.
Activate the Insert ribbon.
Select **Load Family.**
Select the door panel family created at the start of the exercise.

36.
```
Families
  Annotation Symbols
  Generic Models
    Door Panel Mahogany
      Door Panel Mahogany
  Walls
```
You should see the **Door Panel** in the Families area of the Project Browser.

37.
```
Families            Match
  Annotation
  Generic Mo    Type Properties...
    Door Pa
      Doo       Search...
  Walls
```
Right click on the Door Panel and select **Type Properties**.

38.
Glazing	Frosted Glass	
Dimensions		
Width	3' 0 35/64"	
Thickness	0' 2"	
Interior Panel Thickness	0' 1"	
Height	6' 5 49/128"	
Glazing Offset	0' 4"	
Exterior Panel Thickness	0' 1"	

Click on the = next to **Width**.

39.
Existing family parameters of compatible type:

```
<none>
Frame Projection Ext.
Frame Projection Int.
Frame Width
Height
Rough Height
Rough Width
Thickness
Width
```

Select **Width**.

Press **OK**.

This links the panel width to the width of the door opening.

40.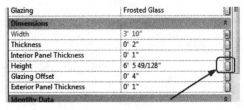
Glazing	Frosted Glass	
Dimensions		
Width	3' 10"	
Thickness	0' 2"	
Interior Panel Thickness	0' 1"	
Height	6' 5 49/128"	
Glazing Offset	0' 4"	
Exterior Panel Thickness	0' 1"	
Identity Data		

Select the = next to **Height**.

41.

Existing family parameters of compatible type:

<none>
Frame Projection Ext.
Frame Projection Int.
Frame Width
Height
Rough Height
Rough Width
Thickness
Width

Select **Height**.

Press **OK**.

42.

Dimensions	
Width	3' 10"
Thickness	0' 2"
Interior Panel Thickness	0' 1"
Height	7' 2"
Glazing Offset	0' 4"
Exterior Panel Thickness	0' 1"

Select the = next to Thickness and link it to the Thickness parameter.
Width, Thickness, and Height should all be grayed out.

43.

Dimensions	
Width	3' 10"
Thickness	0' 2"
Interior Panel Thickness	0' 1"
Height	7' 2"
Glazing Offset	0' 4"
Exterior Panel Thickness	0' 1"

The size of the door panel is now controlled by the Family Type parameters in the door family- the host family.
Press **OK**.

44.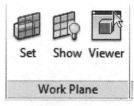

Ref. Level
3D Views
View 1

Activate View 1.

45.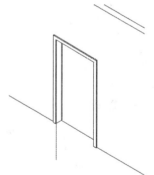

Orient the view so you are looking at the side of the door where the reference line controlling the door swing is located.

46.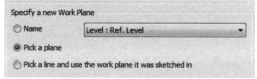

Set Show Viewer

Work Plane

Activate the Create ribbon.

Select **Set Work Plane.**

47.

Specify a new Work Plane
○ Name   Level : Ref. Level
◉ Pick a plane
○ Pick a line and use the work plane it was sketched in

Enable **Pick a plane**.

48.

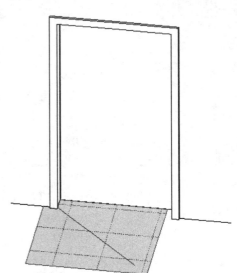

Select the horizontal plane hosted by the reference line.

Use the TAB key to cycle through the selections until the horizontal plane is highlighted.

49.

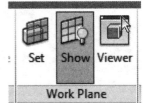

Select **Show Work Plane** to verify that the correct work plane is selected.

50.

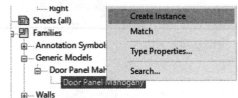

Right click on the Door Panel in the Project Browser.

Select **Create Instance**.

51.

On the ribbon, select **Place on Work Plane**.

52. 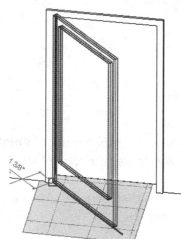 Press the SPACE BAR to rotate the door panel so it is aligned with the reference line.

Locate the ENDPOINT of the reference line to insert the door. This ensures that the door stays constrained to that hinge point.

Right click and select CANCEL to complete placing the door.

53.  Select the door panel.
Select **Edit Work Plane** on the ribbon.

54. The door panel should be hosted by the Reference Line.

Press **OK**.

If the door panel is not hosted by the reference line, then delete the door panel and repeat the steps of setting the work plane and placing the door panel.

55. Select **Family Types**.

56.

Dimensions	
Thickness	0' 2"
Height	7' 2"
Width	3' 10"
Door Swing (default)	15.000°
Rough Width	
Rough Height	

Adjust the door swing and verify that the door moves.
Change the value of the door swing and press **Apply**.

Do this for two to three values to test the family.

57. 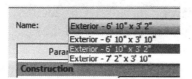 Flex the door sizes and verify that the door panel size changes.

Press **OK**.

58. Launch the **Family Types** dialog.

59. Select **New Parameter**.

60. For Name, type **Exterior Mullion**.
For Type of Parameter, select **Material**.
Group parameter under: **Materials and Finishes**.
Enable Type.
Press **OK**.

Parameter Data
Name:
Exterior Mullion
Discipline:
Common
Type of Parameter:
Material
Group parameter under:
Materials and Finishes

⊙ Type
○ Instance
☐ Reporting Parameter
(Can be used to extract value from a geometric condition and report it in a formula or as a schedulable parameter)

61. Under Materials and Finishes:
Locate the **Exterior Mullion** parameter. Click in the column to assign the material.

Materials and Finishes
Exterior Mullion   <By Category>

62. Select the **Mahogany** material.
Press **OK**.

Project Materials: All
Name
Frosted Glass
Glass
Mahogany
Paint - Canvas

63. You should see the Mahogany material listed.

Materials and Finishes
Exterior Mullion   Mahogany
Dimensions

64. Select **New Parameter**.

65. For Name, type **Interior Mullion**.
For Type of Parameter, select **Material.**
Group parameter under: **Materials and Finishes**.
Enable **Type**.
Press **OK**.

Parameter Data
Name:
Interior Mullion
Discipline:
Common
Type of Parameter:
Material
Group parameter under:
Materials and Finishes

⊙ Type
○ Instance
☐ Reporting Parameter
(Can be used to extract value from a geometric condition and report it in a formula or as a schedulable parameter)

66. 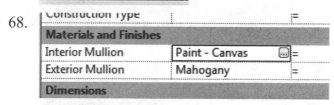 Under Materials and Finishes: Locate the **Interior Mullion** parameter.
Click in the column to assign the material.

67.  Select the **Paint -Canvas** material.
Press **OK**.

68. You should see the Paint-Canvas material listed.
Switch to the other family type sizes and assign the materials.
Close the dialog.

69. 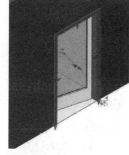 Select the Frame on the Exterior Side.

70.  Set the Material to the **Exterior Mullion**.

Press **OK**.

71. Orbit the view to the interior side.

Select the **Interior Mullion**.

72. 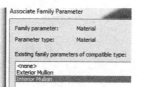 Set the Material to the **Interior Mullion**.

Press **OK**.

73. Materials have to be assigned for each family type. So select each family door size and verify the material has been assigned.

74.  Launch the **Family Types** dialog.

75.  Select **Sort Parameters in Descending Order.**

76. 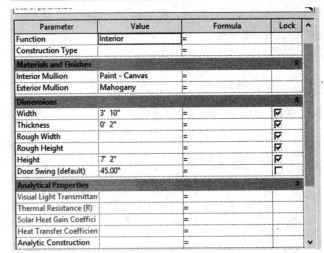 Note how the parameters have re-organized.

Parameter	Value	Formula	Lock
Function	Interior	=	
Construction Type		=	
Materials and Finishes			
Interior Mullion	Paint - Canvas	=	
Exterior Mullion	Mahogany	=	
Dimensions			
Width	3' 10"	=	☑
Thickness	0' 2"	=	☑
Rough Width		=	☑
Rough Height		=	☑
Height	7' 2"	=	☑
Door Swing (default)	45.00°	=	☐
Analytical Properties			
Visual Light Transmittan		=	
Thermal Resistance (R)		=	
Solar Heat Gain Coeffici		=	
Heat Transfer Coefficien		=	
Analytic Construction		=	

77. Save your model as *Door Exterior- Mahogany.rfa.*

78. Open a project file and insert the door and test it.

Exercise 3-6
Creating a Type Catalog

Drawing Name: **Door Exterior- Mahogany.rfa**
Estimated Time to Completion: 55 Minutes

Scope
Create a Type Catalog Using Export.
Modify Using Notepad.
Load into a New Project

Solution

1. Open *Door Exterior- Mahogany.rfa*.

2. 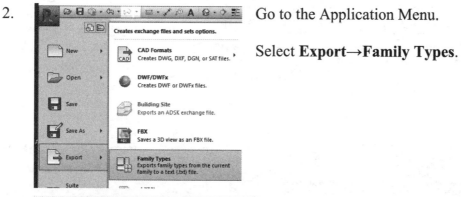 Go to the Application Menu.

   Select **Export→Family Types**.

3. File name: Door Exterior - Mahogany.txt

   Files of type: Text Files (*.txt)

   Browse to your exercise folder.
   Save as a .txt file.
   Press **Save**.

4. Locate the file and open in Notepad.

   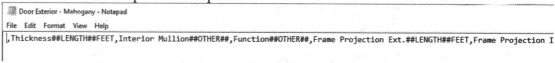

   Door Exterior - Mahogany - Notepad
   File Edit Format View Help
   ,Thickness##LENGTH##FEET,Interior Mullion##OTHER##,Function##OTHER##,Frame Projection Ext.##LENGTH##FEET,Frame Projection I

5. 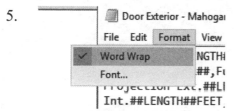 Go to **Format**.
   Enable **Word Wrap**.

6.

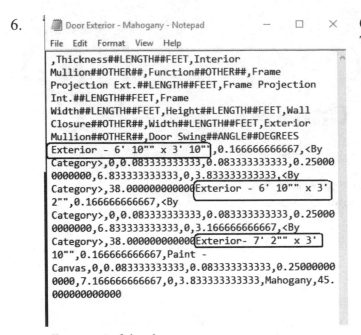

Can you locate the three Door Types in the file?

7.

Highlight one of the type definitions.
Take care to highlight the entire definition.
Press **Ctrl+C** to Copy.

8.

Move your cursor to the end of the text and press **Ctrl+V** to Paste.

9. Change the dimensions to read **80"" x 36""**. Note the double ".

```
Category>,0,0.083333333333,0.083333333333,0.250000000000,6.83333333333
3333333333,<By Category>,38.000000000000Exterior - 80"" x
36"",0.166666666667,<By
Category>,0,0.083333333333,0.083333333333,0.250000000000,6.83333333333
6666666667,<By Category>,38.000000000000
```

10. Change the values for the height and width. The height should be changed to: **6.666666666667** and the width should be changed to **3.000000000000**.

```
3333333333,<By Category>,38.000000000000Exterior - 80"" x
36"",0.166666666667,<By
Category>,0,0.083333333333,0.083333333333,0.250000000000,6.6666666666667,0,3.0
00000000000,<By Category>,38.000000000000
```

11. Save the file and close.

12. 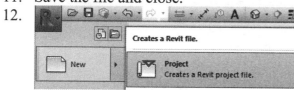 Start a new project using the Architectural template.

13. 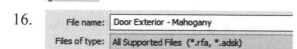 Place a generic wall.

14. Select the **Door** tool.

15. Select **Load Family**.

16. Select the **Door Exterior – Mahogany** family you created.

17. A table now pops up which allows you to select which door sizes you want to load into your project.

Using a Lookup Table allows you to only load the families you plan to use in your project which keeps your project file size smaller.

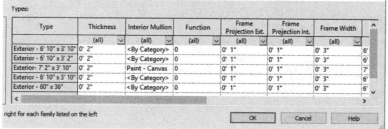

Types:

Type	Thickness	Interior Mullion	Function	Frame Projection Ext.	Frame Projection Int.	Frame Width	
(all)	(all)	(all)	(all)	(all)	(all)	(all)	
Exterior - 6' 10" x 3' 10"	0' 2"	<By Category>	0	0' 1"	0' 1"	0' 3"	6'
Exterior - 6' 10" x 3' 2"	0' 2"	<By Category>	0	0' 1"	0' 1"	0' 3"	6'
Exterior - 7' 2" x 3' 10"	0' 2"	Paint - Canvas	0	0' 1"	0' 1"	0' 3"	7'
Exterior - 6' 10" x 3' 10"	0' 2"	<By Category>	0	0' 1"	0' 1"	0' 3"	6'
Exterior - 80" x 36"	0' 2"	<By Category>	0	0' 1"	0' 1"	0' 3"	6'

right for each family listed on the left   OK   Cancel   Help

18.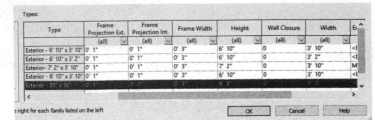

Highlight the new size Exterior- **80" x 36"**.

Press **OK.**

19. Place in the wall.

20.

Switch to a 3D view and inspect your door.

21. Select the door.
Check the Type Selector drop-down.

Only the size which was selected is available.

22. Save as *ex3-5.rvt*.

Exercise 3-7

Assigning OMNI Class Numbers

Drawing Name: **Door-Exterior Mahogany.rvt, Park Bench**
Estimated Time to Completion: 10 Minutes

Scope
Assign an OMNI Class Number to your door family and park bench family

Solution

1. Open the **Door-Exterior Mahogany** family.

2. 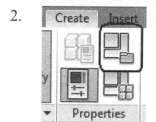 Select the **Family Categories** tool from the Create ribbon.

3. 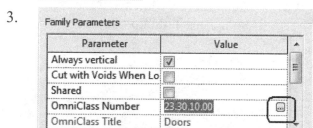 Click on the ... button in the Omni Class Number field.

   Left click in the column to access the ... button.

4. ```
   ⊞ 23.25.00.00 - Structural and Space Division Products
   ⊟ 23.30.00.00 - Openings, Passages, Protection
      ⊟ 23.30.10.00 - Doors
         ⊞ 23.30.10.11 - Door Components
         ⊟ 23.30.10.14 - Passage Doors by Material
              23.30.10.14.11 - Metal Passage Doors
              23.30.10.14.14 - Wood Passage Doors
              23.30.10.14.17 - Plastic Passage Doors
              23.30.10.14.21 - Composite Passage Doors
              23.30.10.14.24 - Glazed Passage Doors
              23.30.10.14.27 - All-Glass Passage Doors
   ```
   Locate the **Glazed Passage Doors** OMNI class.

   Highlight and press **OK**.
   ```

5. 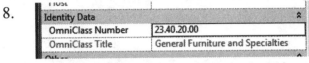 The number updates and the OmniClass Title updates.

   Press **OK**.

6. Save your model.

7. Open the **park bench** family.

8. 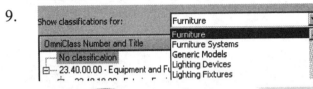 In the Properties pane:
   Click on the ... button in the Omni Class Number field.

9. 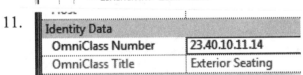 Change the Show Classifications to **Furniture**.

10. 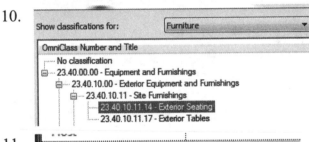 Locate the **Exterior Seating** OMNI class under Site Furnishings.

    Highlight and press **OK**.

11. | Identity Data | |
    | --- | --- |
    | OmniClass Number | 23.40.10.11.14 |
    | OmniClass Title | Exterior Seating |

    The number updates and the OmniClass Title updates.

    Press **OK**.

12. Save your model.

OMNI Classes are used for scheduling and keynotes.

Exercise 3-8

Room Calculation Point

Drawing Name: **i_room_calculation_point**
Estimated Time to Completion: 20 Minutes

Scope
Modify a room calculation point on a family so it associates with the correct room in a schedule

Solution

1.

   ```
   ···· Legends
   ]···  Schedules/Quantities
         ···· Furniture Schedule
         ···· Plumbing Fixture Schedule
         ···· Specialty Equipment Schedule
   ```

   Open the **Furniture Schedule**.

2.

<Furniture Schedule>			
A	B	C	D
Family and Type	Count	Room: Name	Level
Corbu armchair: Corbu armchair	1	LIVING	1st Floor
Corbu armchair: Corbu armchair	1	LIVING	1st Floor
Corbu armchair: Corbu armchair	1	LIVING	1st Floor
Corbu armchair: Corbu armchair	1	LIVING	1st Floor
Corbu armchair: Corbu armchair	1		1st Floor

   There is one piece of furniture not associated with a room.

3. 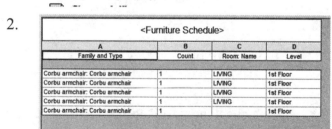   Highlight the chair with no room shown.

4.

   Highlight in Model
   Element

   Select **Highlight in Model** on the ribbon.

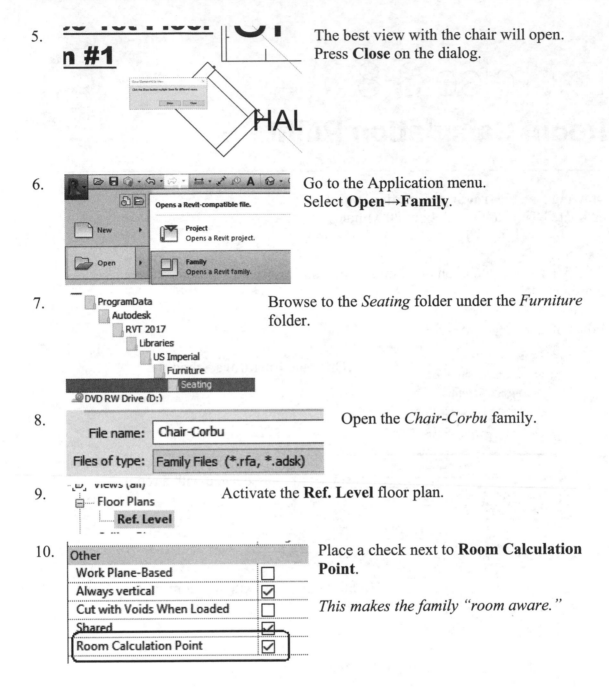

5. The best view with the chair will open. Press **Close** on the dialog.

6. Go to the Application menu. Select **Open→Family**.

7. Browse to the *Seating* folder under the *Furniture* folder.

8. Open the *Chair-Corbu* family.

9. Activate the **Ref. Level** floor plan.

10. Place a check next to **Room Calculation Point**.

This makes the family "room aware."

11. 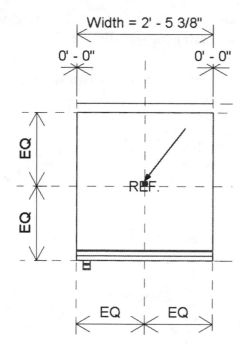 A green dot will appear to indicate the location of the Room Calculation Point.

12. 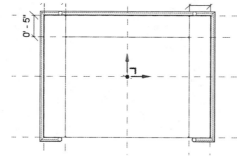 Select the Room Calculation Point.
A gizmo will appear when it is selected.

13. 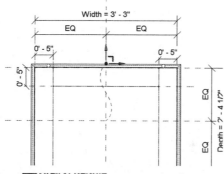 Use the green vertical arrow to move the Room Calculation Point to the middle of the back of the chair.

14. 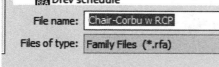 Save the family to your exercise folder with a new name: **Chair- Corbu w RCP**.

15. Load the family into the open project.

16.

Select the chair. Right click and **Select All Instances→Visible in View**.

17.
| Search |
| Chair-Corbu w RCP |
| Chair |

Use the Type Selector to select the **Chair- Corbu w RCP.**

The chair will shift position based on the new RCP.

18.
- Legends
- Schedules/Quantities
  - **Furniture Schedule**
  - Plumbing Fixture Schedule
  - Specialty Equipment Schedule

Open the **Furniture Schedule**

19.

<Furniture Schedule>

A	B	C	D
Family and Type	Count	Room	Level
Chair-Corbu w RCP: Chair	1	LIVING	1st Floor
Chair-Corbu w RCP: Chair	1	LIVING	1st Floor
Chair-Corbu w RCP: Chair	1	LIVING	1st Floor
Chair-Corbu w RCP: Chair	1	LIVING	1st Floor
Chair-Corbu w RCP: Chair	1	LIVING	1st Floor

Note the chair is now associated with a room.

Close without saving.

Certified User Practice Exam

1. If a level-based component is moved from Level 1 to Level 2 by changing the Element Properties, the new location is:

    A. At the origin.
    B. Corresponds to the location on the previous level.
    C. Where the user selects it to be.
    D. None of the above.

2. Families in the Project Browser are organized by:

    A. Instance, then Category, then Family
    B. Family, then Instance
    C. Category, then Family, then Type
    D. Family, then Type

3. A _____ is a group of elements with a common set of properties (called parameters) and a related graphical representation.

    A. view
    B. family
    C. project
    D. model
    E. mass

4. Select TWO characteristics you can specify for a family type:

    A. Category
    B. Materials
    C. Levels
    D. Views
    E. Dimensions

5. Select the THREE types of families:

    A. Datum
    B. View
    C. System
    D. Model
    E. Mass

6. A(n) _____ family exists only in the project and can not be loaded into other projects.

   A. massing
   B. system
   C. loadable
   D. in-place
   E. model

7. Select the three items which are system families:

   A. Walls
   B. Doors
   C. Roofs
   D. Lighting Fixtures
   E. Ceilings

8. Levels, grids, sheets, and viewports are:

   A. Loadable families
   B. System families
   C. In-place families
   D. Project-based

Answers:
 1) B; 2) C; 3) B; 4) B & E; 5) C, D, & E; 6) B; 7) A, C, & E; 8) B

Certified Professional Practice Exam

1. To change or assign the OMNI class of a model family, select:

   A. Family Types
   B. Family Category and Parameters
   C. Properties
   D. Materials

2. To create a new system family you:

   A. Use a Revit family template
   B. Modify the instance parameters of an existing similar family
   C. Duplicate a similar system family and modify the type parameters
   D. Modify the type parameters of a similar system family.

3. In the Generic Model family template, you are provided with:

   A. Levels
   B. Reference Planes
   C. Grids
   D. Grid Guides

4. To create an opening in an extrusion when defining a model family, you need to place a:

   A. Opening
   B. Void
   C. Hole
   D. Cut

5. When a room schedule does not reflect the correct room for an element, the element family should be modified by:

   A. Moving it into the correct room
   B. Modifying the element's base point
   C. Making the element "room-aware"
   D. Modifying the room schedule

Answers:
   1) B or C; 2) C; 3) B; 4) B; 5) C

View Properties

This lesson addresses the following exam questions:

- View Properties
- Object Visibility Settings
- Section Views
- Elevation Views
- View Templates
- Scope Boxes

The Project Browser lists all the views available in the project. Any view can be dragged and dropped onto a sheet. Once a view is used or consumed on a sheet, it cannot be placed a second time on a sheet – even on a different sheet. Instead, you must create a duplicate view. You can create as many duplicate views as you like. Each duplicate view may have different annotations, line weight settings, detail levels, etc. Annotations are specific to a view. If a view is deleted, any annotations are also deleted.

Revit has bidirectional associativity. This means that changes in one view are automatically reflected in all views. For example, if you modify the dimensions or locations of a window in one view, the change is reflected in all the views, including the 3D view.

You can control the appearance of Revit elements using Object Visibility Settings. These settings control line color, line type, and line weight. You can create templates which have different Object Visibility Settings for different project types.

Exercise 4-1
Creating a Level

Drawing Name: **i_levels.rvt**
Estimated Time to Completion: 5 Minutes

Scope
Placing a level.

Solution

1.    Activate the **South Elevation**.

   The level names have been turned off.

2. 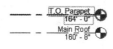   Select each level and place a check in the square that appears. This will turn on visibility of the level name.

3.   You should be able to identify the names for each level.

4. Select the **Level** tool from the Architecture ribbon.

5. 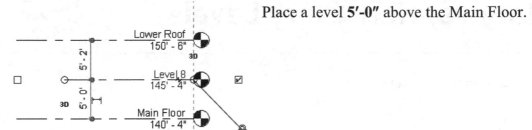 Place a level **5′-0″** above the Main Floor.

6. 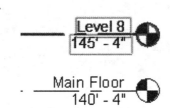 Note the elevation value for the new level.

7. Close without saving.

Exercise 4-2
Story vs. Non-Story Levels

Drawing Name: **story_levels.rvt**
Estimated Time to Completion: 15 Minutes

Scope
Understanding the difference between story and non-story levels
Converting a non-story level to a story level.

Solution

1. Activate the **South Elevation**.

2. Select each level and place a check in the square that appears. This will turn on visibility of the level name.

3. Study the Main Floor level. Notice that it is the color black while all the other levels are blue.

   The Main Floor level is a non-story or reference level. It does not have a view associated with it.

4. Activate the Architecture ribbon.
   Select the **Level** tool on the Datum panel.

5. On the Options bar: Uncheck **Make Plan View**. Set the Offset to **8' 0"**.

6. Select the **Pick** tool on the Draw panel.

7. 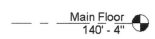 Select the Main Floor level.
Verify that the preview shows the level will be placed 8' 0" ABOVE the Main Floor level.

8. 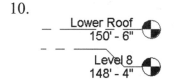 The level is placed above the Main Floor.
Right click and select Cancel twice to exit the Level command.

9. Select the Elbow control on the new level to add a jog.

10. Note that the new level is also a non-story or reference level.
Check in the Project Browser and you will see that no views were created with the new level.

11. 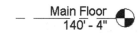 Activate the View ribbon.
Select the **Plan Views→Floor Plan** tool on the Create panel.

12. 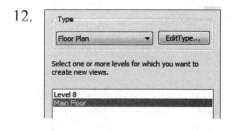 The reference levels are listed.
Select the **Main Floor** level and press **OK**.

13.
Floor Plans
    Ground Floor
    Lower Roof
    Main Floor
    Main Roof
    Site
    T.O. Footing
    T.O. Parapet

The Main Floor floor plan view will open.
Note that the Main Floor floor plan is now listed in the Project Browser; however, there is no ceiling plan for the Main Floor.

14.

Plan Views ▾   Draftin
   Floor Plan
   Reflected Ceiling Plan

Activate the View ribbon.
Select the **Plan Views→Reflected Ceiling Plan** tool on the Create panel.

15.

Type
Ceiling Plan ▾   EditType...

Select one or more levels for which you want to create new views.

Level 8
Main Floor

The reference levels are listed.
Select the **Main Floor** level and press **OK**.

16.
Ceiling Plans
    Ground Floor
    Lower Roof
    Main Floor
    Main Roof
    T.O. Footing
    T.O. Parapet

The Main Floor ceiling plan view will open.

17.
T.O. Parapet
Elevations (Building Elevation)
    East
    North
    South
    West

Activate the **South Elevation**.

18.
150' - 6"
Level 8
148' - 4"

Main Floor
140' - 4"

The Main Floor level is now the color Blue to indicate it has story views.

19. Close the file without saving.

Exercise 4-3
Creating Column Grids

Drawing Name: **i_grids.rvt**
Estimated Time to Completion: 10 Minutes

Scope
Placing column grids.

Solution

1.  Activate the **T.O. Footing** view.

2. 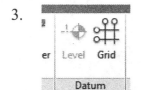 Zoom into the building area displayed.

3. Select the **Grid** tool on the Datum panel from the Architecture ribbon.

4. Select the **Pick Lines** mode from the Draw panel.

5.  Set the Offset to **2'-0″** [**600 mm**] on the Options bar.

6. 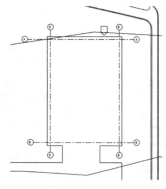 Click to place a vertical grid line as shown.

7. 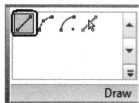 Place gridlines as shown.

   Use the grips by the heads to drag the grid bubbles into position.

8. 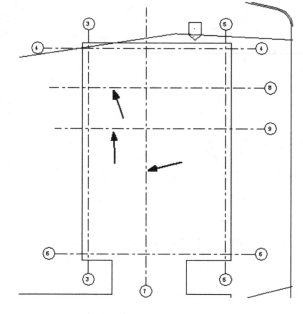 Switch to draw grid mode.

   Add a vertical grid line as shown.

   Add two horizontal grid lines as shown.

9. 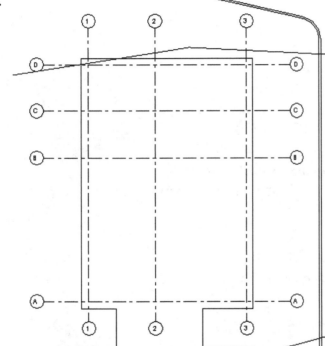 Place a check on both rectangles to make bubbles visible on both sides of the grid line.

10. Label the horizontal grid lines A-D.
The alpha labeled grids are incremented from bottom to top.

Label the vertical grid lines 1-3.
The number labeled grids are incremented from left to right.

Enable the grid bubbles on both ends.

Turn off the visibility of elevations.

11.  Activate the **Annotate** ribbon.
Select the **ALIGNED** dimension tool from the Dimension panel.

12. 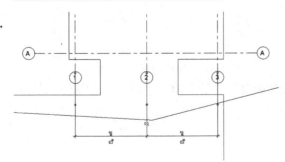 Place a multi-segmented dimension between the three vertical grid lines.
Enable the EQ toggle.

13. Place a multi-segmented dimension between the horizontal grid lines.

Enable the EQ toggle.

To place a continuous dimension, select each grid line in order, then left click above (if the last grid line is the top one) or below (if the last grid line is the bottom one).

Exit the command.

14. Select the vertical dimensions.
Right click and toggle **EQ Display** off.

15. The dimension is now visible.

16. Select the horizontal dimensions.
Right click and toggle **EQ Display** off.

The horizontal dimensions are now visible.

17. Activate the Architecture ribbon.
Select the **Structural Column** tool from the Build panel.

18. 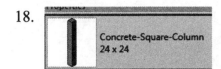 Select the **24 x 24 Concrete Square Column [600 x 600 mm]** from the Type Selector list on the Properties pane.

19. On the Options bar: Select **Height** from the drop-down. Select **Main Floor** to set the Height.

20. 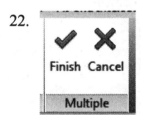 Enable the **At Grids** option to place columns at grid intersections.

21. 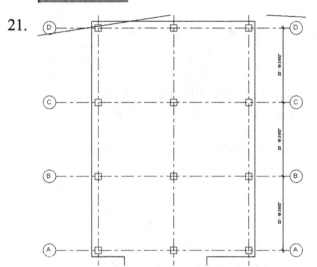 Hold down the CONTROL key.
Select each grid line.

A column will be placed at each grid intersection.

22. Select the Green Check when you see a column at every grid intersection.
Right click and select CANCEL to exit the command.

23. Activate the Annotate ribbon.
Select **Tag All** from the Tag panel.

24.

Category	Loaded Tags
Door Tags	Door Tag
Room Tags	Room Tag
Room Tags	Room Tag : Room Tag With Area
Structural Column Tags	Structural Column Tag
Wall Tags	Wall Tag : 1/2"
Wall Tags	Wall Tag : 1/4"
Window Tags	Window Tag

◉ All objects in current view
○ Only selected objects in current view
☐ Include elements from linked files

Highlight the **Structural Column Tag**.

Press **OK**.

25.

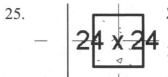

Zoom into a column to read the tag.

The column is labeled with the Column Type.

26.

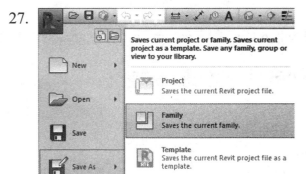

Cancel

Repeat [Tag All Not Tagged]
Recent Commands

Hide in View
Override Graphics in View

Create Similar
Edit Family
Select Previous

Select one of the structural column tags.
Right click and select **Edit Family**.

Make sure you select the tag and not the column.

27.

Saves current project or family. Saves current
project as a template. Save any family, group or
view to your library.

New

Project
Saves the current Revit project file.

Open

Family
Saves the current family.

Save

Template
Saves the current Revit project file as a
template.

Save As

Go to **File→Save As→Family**.

28.

File name: Structural Column Location Tag.rfa
Files of type: Family Files (*.rfa)

Rename the tag *Structural Column Location Tag.rfa*.

29.

Other (1)   ▼ Edit Type

Graphics   ⌃
Sample Text   1i
Label   [Edit...]
Wrap between parameter... ☐
Vertical Align   Middle

Select the text.
Select the **Edit** button in the Label field on the Properties pane.

30. Use the Add and Remove tools in the middle of the dialog to remove the Type Name and add the Column Location Mark.

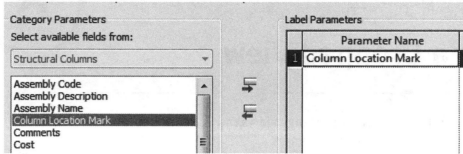

31. Press **OK**.
32. Save the file.
33. 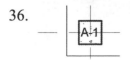 Select the **Load into Project** tool.

    Select the project with the columns.

34. 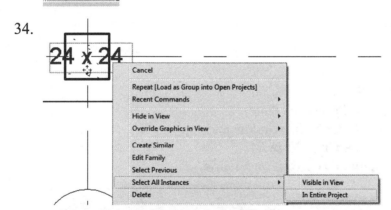 Select one of the column tags.

    Right click and select **Select All Instances→ Visible in View**.

35. 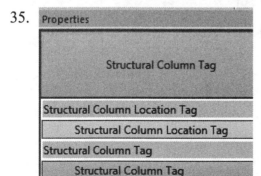 Select the **Structural Column Location Tag** from the Type Selector on the Properties pane.

36. The tags update with the location.

37. Close without saving.

Exercise 4-4

Create a Cropped View

Drawing Name: **i_firestation_managing_views.rvt**
Estimated Time to Completion: 10 Minutes

Scope
Create a cropped view

Solution

1.
   - Floor Plans
     - Ground Floor
     - Lower Roof
     - Main Floor
     - **Main Floor- Furniture Plan**
     - Main Roof
     - Site
     - T. O. Footing
     - T. O. Parapet

   Activate the **Main Floor – Furniture Plan** view.

2.
Referencing Detail	
Extents	
Crop View	☑
Crop Region Visible	☑
Annotation Crop	☐

   In the Properties Pane:
   Scroll down the window and:
   Enable **Crop View**.
   Enable **Crop Region Visible**.

4.

   Zoom out.

   Select the viewport rectangle.

5.

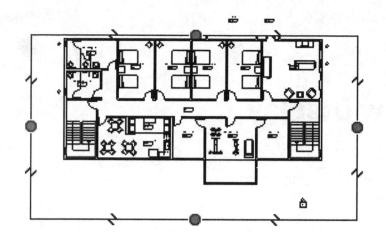

Use the bubbles to position the viewport so that only the furniture floor plan is visible.

6. Select **Hide Crop Region** using the tool in the View Control bar. Press **OK**.

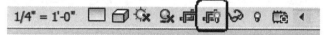

7. Close without saving.

Exercise 4-5
Change View Display

Drawing Name: **i_firestation_managing_views.rvt**
Estimated Time to Completion: 15 Minutes

Scope
Use Temporary Hide/Isolate to control visibility of elements.
Change Line Width Display
Change Object Display Settings

Solution

1.
Floor Plans
  Ground Floor
  Lower Roof
  Main Floor
  Main Floor- Furniture Plan
  Main Roof
  Site

Activate the **Main Floor** floor plan.

2.
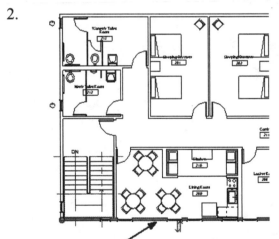

Select one of the exterior walls so it is highlighted.

3.
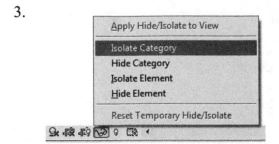

Select the **Temporary Hide/Isolate** tool.
Right click and select **Isolate Category**.

4-16

4.

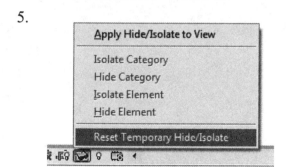

Only the exterior walls are visible.

5.

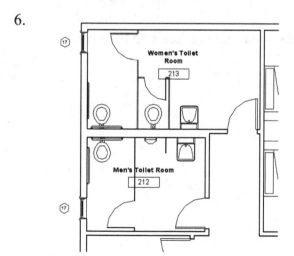

Select the **Temporary Hide/Isolate** tool. Right click and select **Reset Temporary Hide/Isolate**.

6.

Zoom into the region where the lavatories are located.

7. 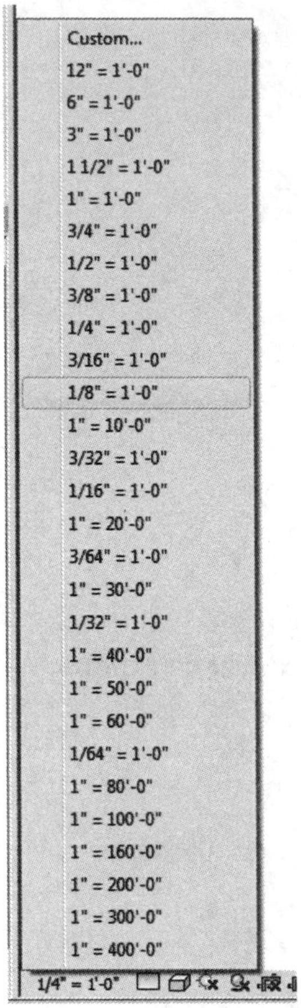
Custom...
12" = 1'-0"
6" = 1'-0"
3" = 1'-0"
1 1/2" = 1'-0"
1" = 1'-0"
3/4" = 1'-0"
1/2" = 1'-0"
3/8" = 1'-0"
1/4" = 1'-0"
3/16" = 1'-0"
1/8" = 1'-0"
1" = 10'-0"
3/32" = 1'-0"
1/16" = 1'-0"
1" = 20'-0"
3/64" = 1'-0"
1" = 30'-0"
1/32" = 1'-0"
1" = 40'-0"
1" = 50'-0"
1" = 60'-0"
1/64" = 1'-0"
1" = 80'-0"
1" = 100'-0"
1" = 160'-0"
1" = 200'-0"
1" = 300'-0"
1" = 400'-0"
1/4" = 1'-0"

Change the view scale to **1/8″ = 1′-0″**.

8. 
Women's Toilet Room
213
Men's Toilet Room
212
17
17

Note that the room tags scale to the view.

9. 
Modify

Activate the **Modify** ribbon.

10. 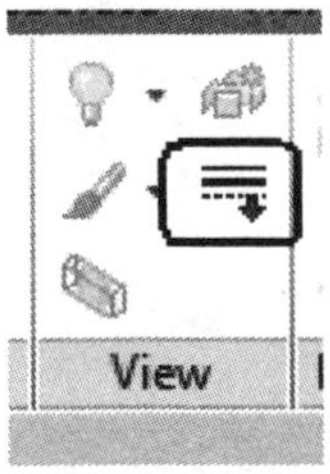
View

Select the **Linework** tool on the View panel.

11. Set the Line Style to **Overhead**.

12. 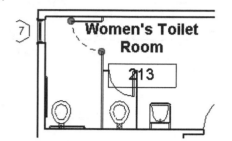 Select the door swing on the toilet cubicle.

Note that the door swing's appearance changes.

13. Activate **Section 1**.

14. Activate the Manage ribbon.

Select **Settings→Object Styles**.

15.

Expand the **Doors** category on the Model Objects tab.
Change the Line Weight, Line Color, and Line Pattern for the Panel and Frame.
Press **Apply** to see the changes.

16. 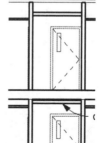 You can move the dialog over so you can see how the display is changed.

Press OK to close the dialog.

Note that linework changes are specific to the view, but object settings changes affect all views.

17. Close without saving.

Exercise 4-6
Duplicating Views

Drawing Name: **duplicating_views.rvt**
Estimated Time to Completion: 20 Minutes

Scope
Duplicate view with Detailing
Duplicate view as Dependent
Duplicate view
Understand the difference between the different duplicating views options

Solution

1. 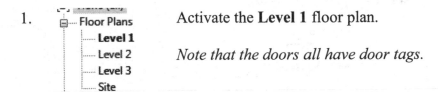 Activate the **Level 1** floor plan.

   Note that the doors all have door tags.

2. 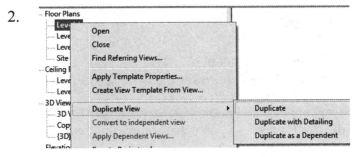 Highlight Level 1.
   Select **Duplicate View→Duplicate**.

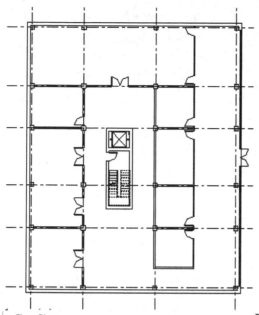

When you select Duplicate View→Duplicate, any annotations, such as tags and dimensions, are not duplicated.

3.

Highlight the copied level 1.

Right click and select **Rename**.

Name: Level 1 - No Annotations

Type **Level 1- No Annotations**.

Press **OK**.

4.

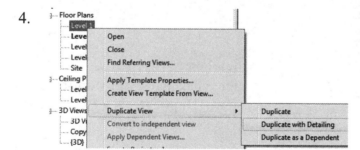

Highlight Level 1.
Right click and select **Duplicate View→Duplicate with Detailing**.

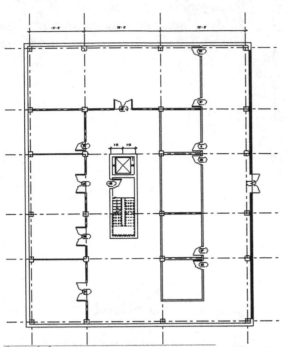

Note that when you select Duplicate with Detailing the new view includes annotations, such as tags and dimensions.

5.

Highlight the copied level 1.

Right click and select **Rename**.

Name: Level 1-Original Annotations

Type **Level 1- Original Annotations**.

Press **OK**.

6.

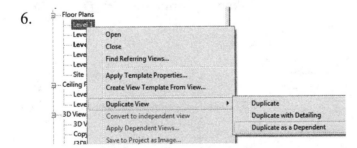

Highlight Level 1.
Right click and select **Duplicate View→Duplicate as a Dependent**.

7.

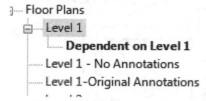

Notice that the duplicated view includes the annotations.

- ┆— Floor Plans
  - └— Level 1
    - └— **Dependent on Level 1**
    - ┄— Level 1 - No Annotations
    - ┄— Level 1-Original Annotations

In the Project Browser, the dependent view is listed underneath the parent view.

8.

Name: | Level 1 - Stairs Area|

Rename the dependent view **Level 1- Stairs area**.

9.

- ┆— Floor Plans
  - ┆— Level 1
    - **Level 1 - Stairs Area**
    - Level 1 - No Annotations

Activate **Level 1- Stairs area**.

10.

Extents	
Crop View	☑
Crop Region Vis...	☑
Annotation Crop	☑

In the Properties pane:
Enable **Crop View**.
Enable **Crop Region Visible**.

11.

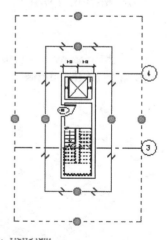

Adjust the crop region to focus the view on the stairs area.

12.

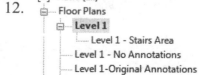

- ┆— Views (all)
  - ┆— Floor Plans
    - ┆— **Level 1**
      - └— Level 1 - Stairs Area
    - ┄— Level 1 - No Annotations
    - ┄— Level 1-Original Annotations

Activate **Level 1**.

13. Zoom into the stairs area.

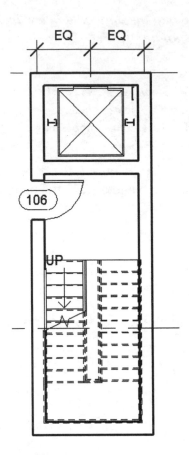

14. Annotate   Activate the **Annotate** ribbon.

15. Select the **Tread Number** tool on the Tag panel.

Tread
Number

Tread Numbers can only be applied to Component-based stairs.

16. 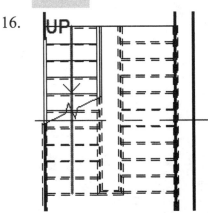 Select the middle line that highlights when your mouse hovers over the left side of the stairs.

17.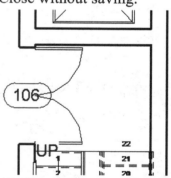

Select the middle line on the right side of the stairs. Right click and select CANCEL to exit the command.

18.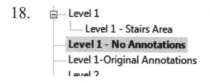

Activate **Level 1- Original Annotations**.

Notice that the new annotations - the tread numbers - are not visible in this view.

19.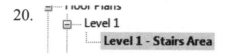

Activate **Level 1 - No Annotations**.

Notice that the new annotations - the tread numbers - are not visible in this view.

20.

Activate the **Level 1- Stairs area** dependent view.

Notice that any annotations added to the parent view are added to the dependent view.

21. Close without saving.

Extra: Change the door labeled 106 to a double flush door. Which Level 1 views display the new door type?

Exercise 4-7

Setting View Depth

Drawing Name: **i_firestation_basic_plan.rvt**
Estimated Time to Completion: 5 Minutes

Scope
Determine the view depth of a view

Solution

1. 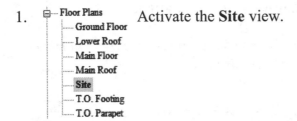 Activate the **Site** view.

2. 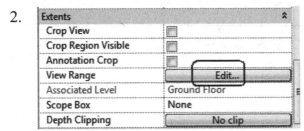 In the Properties pane:
   Scroll down to **View Range** located under the Extents category.
   Select the **Edit** button.

3. 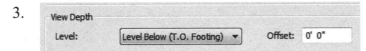 Determine the **View Depth**.

4. Press **OK**.

5. Close without saving.

Exercise 4-8

Reveal Hidden Elements

Drawing Name: **i_visibility.rvt**
Estimated Time to Completion: 5 Minutes

Scope
Turn on the display of hidden elements

Solution

1.
   - Floor Plans
     - Ground Floor
     - **Ground Floor Admin Wing**
     - Lower Roof
     - Main Floor
     - Main Roof
     - Site
     - T. O. Footing
     - T. O. Parapet

   Activate **the Ground Floor Admin Wing** floor plan.

2.   1" = 10'-0"   Select the **Reveal Hidden Elements** tool.

3.   Items highlighted in magenta are hidden. Window around the two tables while holding down the CONTROL key to select them.

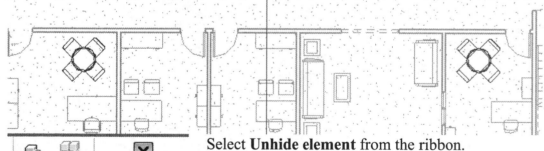

4.   Select **Unhide element** from the ribbon.

Unhide Element Unhide Category Toggle Reveal Hidden Elements Mode
Reveal Hidden Elements

5. The tables will no longer be displayed as magenta (hidden elements).

6. Select the **Close Hidden Elements** tool.

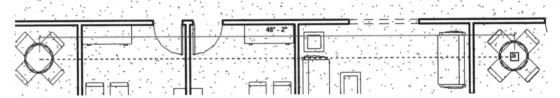

7. The view will be restored.

8. Select the **Measure** tool from the Quick Access toolbar.

9. Determine the distance between the center of the two tables.
    Did you get 48' 2"?

    If you didn't get that measurement, check that you selected the midpoint or center of the two tables.

10. Close without saving.

Exercise 4-9

Create a View Template

Drawing Name: **view_templates.rvt**
Estimated Time to Completion: 30 Minutes

Scope
Apply a wall tag
Create a view template
Create a view filter
Apply view settings to a view

Solution

1. Floor Plans
   Level 1
   Level 2
   Level 3
   Site

   Activate Level 1.

2. Annotate   Activate the **Annotate** ribbon.

3. Select **Tag All**.

   Tag
   All

4.

Structural Framing Tags	Structural Framing Tag : Standard
Wall Tags	Wall Tag : 1/2"
Wall Tags	Wall Tag : 1/4"
Wall Tags	wall tag- fire rating
Window Tags	Window Tag

Highlight the **Wall tag - fire rating** as the tag to be used and press **OK**.

The wall tag - fire rating is a custom family. It was pre-loaded into this exercise, but is included with the exercises files on the publisher's website for your use.

5. Zoom in to inspect the tags.

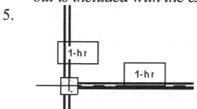

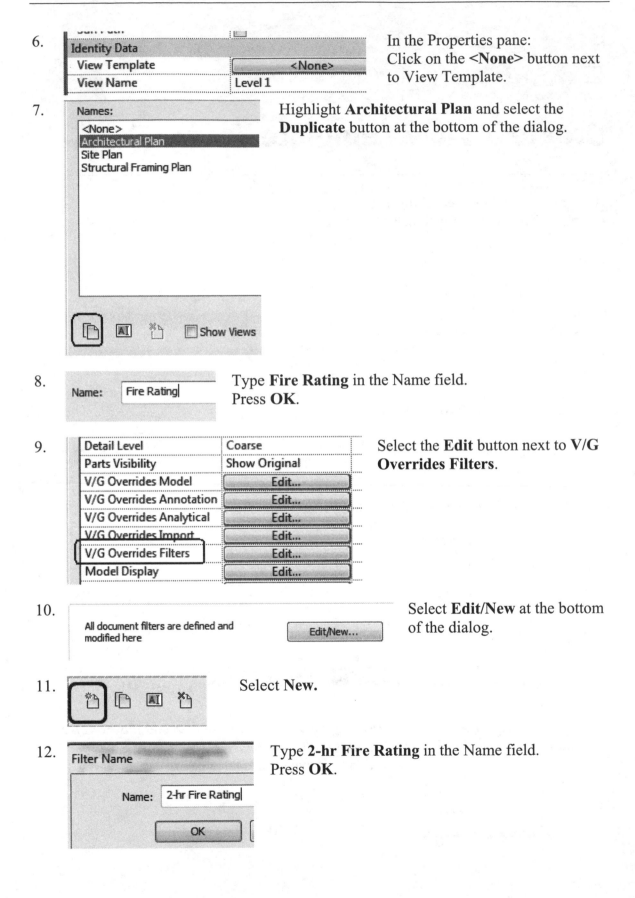

6. In the Properties pane:
Click on the **<None>** button next to View Template.

7. Highlight **Architectural Plan** and select the **Duplicate** button at the bottom of the dialog.

8. Type **Fire Rating** in the Name field.
Press **OK**.

9. Select the **Edit** button next to **V/G Overrides Filters**.

10. Select **Edit/New** at the bottom of the dialog.

11. Select **New.**

12. Type **2-hr Fire Rating** in the Name field.
Press **OK**.

13. 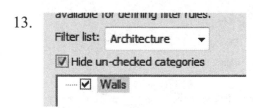 Place a check next to **Walls**.
Then place a check next to **Hide un-checked categories**.

14. 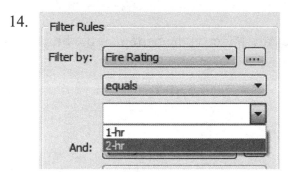 Set the Filter Rules to
Filter by: Fire Rating
Equals
2-hr.

Press **Apply**.

Apply saves the changes and keeps the dialog open. OK saves the changes and closes the dialog.

15. 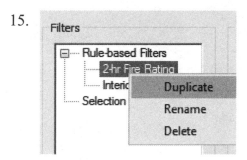 Right click on the **2-hr Fire Rating** filter.
Select **Duplicate**.

16. 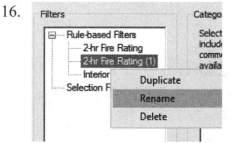 Highlight the copied filter.
Right click and select **Rename**.

17.  Rename to **1-hr Fire Rating**.
Press **OK**.

18. 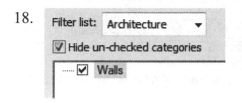 Place a check next to **Walls**.
Then place a check next to **Hide un-checked categories**.

19.

Filter by: Fire Rating

equals

1-hr

Set the Filter Rules to
Filter by: Fire Rating
Equals
1-hr.
Press **Apply**. Press OK.

20.

Name	Vi

No filters have been applied to this vi

Add   Remove

Select the **Add** button at the bottom of the Filters tab.

21.

Select one or more filters to insert.

Rule-based Filters
    1-hr Fire Rating
    2-hr Fire Rating
    Interior
Selection Filters

Hold down the Control key.
Highlight the 1-hr and 2-hr fire rating filters and press **OK**.

22.

| Model Categories | Annotation Categories | Analytical Model Categories | Imported Categories | Filters |

Name	Visibility	Projection/Surface			Cut		Halftone
		Lines	Patterns	Transparen...	Lines	Patterns	
2-hr Fire Rating	✓	Override...	Override...	Override...	Override...	Override...	☐
1-hr Fire Rating	✓						☐

You should see the two fire rating filters listed.

If the interior filter was accidentally added, simply highlight it and select Remove to delete it.

23.

Name	Visibility	Projection/Surface		
		Lines	Patterns	Tr:
2-hr Fire Rating	☑	Override...	Override...	C
1-hr Fire Rating	☑			

Highlight the **2-hr Fire Rating** filter.

24. Select the **Pattern Override** under Projection/Surface.

25.

Pattern: [2 Hour ▼]

Select the **2 Hour** fill pattern. Press **OK.**

26.

Pattern Overrides

☑ Visible

Color: [Blue]

Pattern: [2 Hour ▼] [...]

Set the Color to **Blue.** Press **OK.**

27.

Name	Visibility	Projection/Surface		
		Lines	Patterns	Tr:
2-hr Fire Rating	☑			
1-hr Fire Rating	☑	Override...	Override...	C

Highlight the **1-hr Fire Rating** filter

28. Select the **Pattern Override** under Projection/Surface.

29.

Pattern Overrides

☑ Visible

Color: [Magenta]

Pattern: [1 Hour ▼] [...]

Set the fill pattern to **1 Hour** and the Color to **Magenta.** Press **OK.**

30. Apply the same settings to the Cut Overrides. Press **OK.**

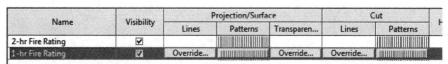

Name	Visibility	Projection/Surface			Cut		
		Lines	Patterns	Transparen...	Lines	Patterns	H
2-hr Fire Rating	☑						
1-hr Fire Rating	☑	Override...		Override...	Override...		

31.

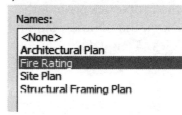

Names:

<None>
Architectural Plan
Fire Rating
Site Plan
Structural Framing Plan

Highlight the **Fire Rating** View Template. Press **Apply.** Press **OK.**

32. 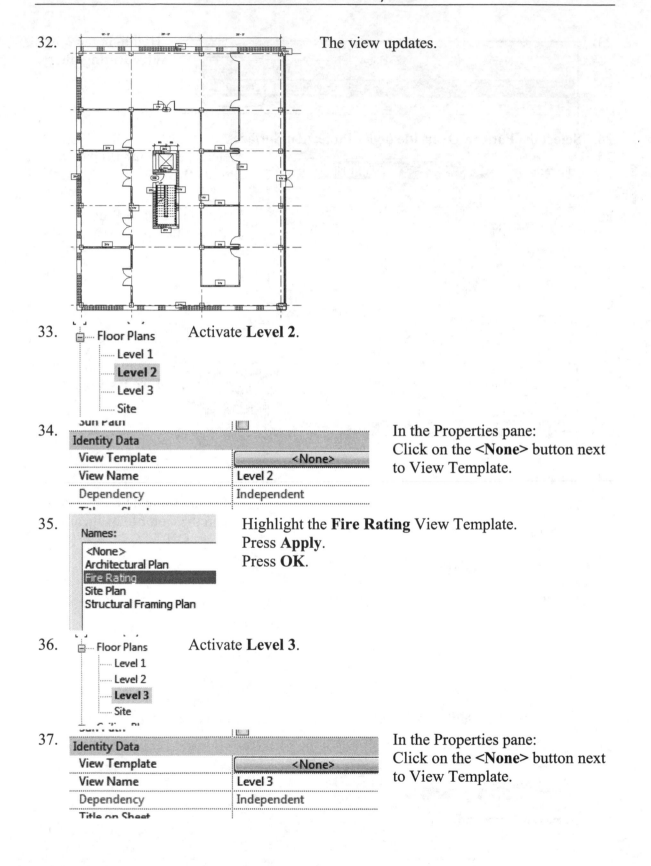 The view updates.

33. Activate **Level 2**.

34. In the Properties pane:
Click on the **<None>** button next to View Template.

35. Highlight the **Fire Rating** View Template.
Press **Apply**.
Press **OK**.

36. Activate **Level 3**.

37. In the Properties pane:
Click on the **<None>** button next to View Template.

38. 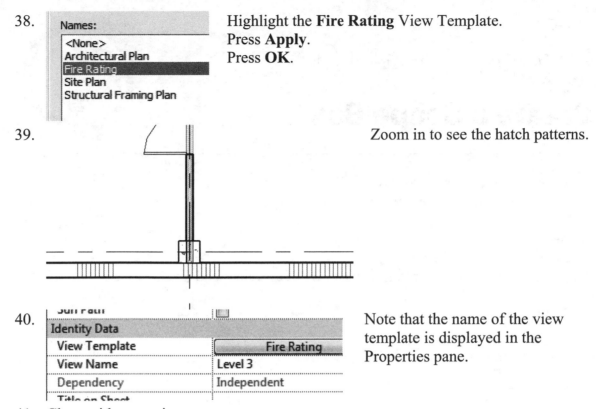 Highlight the **Fire Rating** View Template.
Press **Apply**.
Press **OK**.

39. Zoom in to see the hatch patterns.

40. Note that the name of the view template is displayed in the Properties pane.

41. Close without saving.

The 1-hr, 2-hr, and 3-hr hatch patterns are custom fill patterns provided inside this exercise file. The *.pat files are included with the exercise files on the publisher's website for your use.

Exercise 4-10
Create a Scope Box

Drawing Name: **i_scope_box.rvt**
Estimated Time to Completion: 15 Minutes

Scope
Create and apply a scope box.
Scope boxes are used to control the visibility of grid lines and levels in views.

Solution

1. 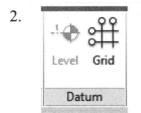   Activate the **Level 1** floor plan.

2. 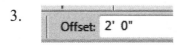   Activate the **Architecture** ribbon.
   Select the **Grid** tool from the Datum panel.

3. 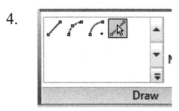   Set the Offset to **2' 0"** on the Options bar.

4.   Select the **Pick Lines** tool from the Draw panel.

5.

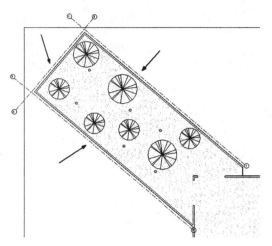

Place three grid lines using the exterior side of the walls to offset.

6.

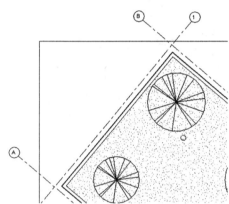

Re-label the grid bubbles so that the two long grid lines are A and B and the short grid line is 1.

7. Select the **Measure** tool from the Quick Access toolbar.

8. Measure one side of the yard area.

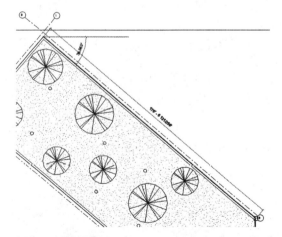

9.

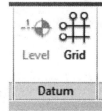

Activate the **Architecture** ribbon.
Select the **Grid** tool from the Datum panel.

10. Offset: 180' 0"   Set the Offset to **180′ 0″** on the Options bar.

11. Draw   Select the **Pick Lines** tool from the Draw panel.

12. 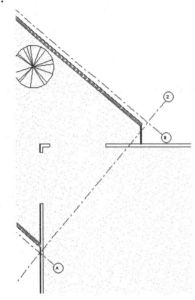   Place the lower grid line by selecting the upper wall and offsetting 180′.
Re-label the grid line **2**.

13. Plan Views ▾ Drafting View Schedules ▾
    Elevation ▾ Duplicate View ▾ Scope Box
    Legends ▾
    Create

    Activate the **View** ribbon.
    Select the **Scope Box** tool from the Create panel.

14. Place the scope box.

    Use the Rotate icon on the corner to rotate the scope box into position.

    Use the blue grips to control the size of the scope box.

15. 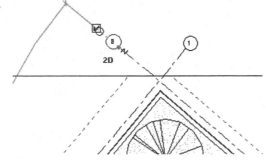 Select the grid line labeled **B**.

16.

Grids (1)	
Identity Data	
Name	B
Extents	
Scope Box	Scope Box 1

In the Properties pane:
Set the Scope Box to Scope Box 1, the scope box which was just placed.

17. Repeat for the other three grid lines.

18.

Scope Boxes (1)	
Identity Data	
Name	Scope Box 1
Extents	
Views Visible	Edit...

Select the Scope Box.
Select **Edit** next to Views Visible in the Properties pane.

19. Click on the column headers to change the sort order.

View Type	View Name	Automatic visibility	Override
3D View	{3D}	Visible	None
Ceiling Plan	Level 1	Visible	None
Ceiling Plan	Level 2	Visible	None
Elevation	South Elevation	Invisible	None
Floor Plan	Level 1	Visible	None
Floor Plan	Level 2	Visible	Invisible
Floor Plan	Site	Visible	None
Floor Plan	Level 3	Visible	None

Set the Level 2 Floor Plan Override Invisible.
Press **OK**.

20.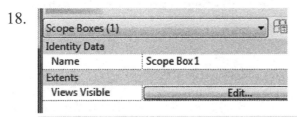

Select the scope box.
Right click and select **Hide in View→ Elements**.

The scope box is no longer visible in the view.

21. Floor Plans
    ├── Level 1
    ├── **Level 2**
    └── Level 3

Activate Level 2.

The grid lines and scope box are not visible.

22. Close the file without saving.

Tip: To make the hide/isolate mode permanent to the view; on the View Control bar, click the glasses icon and then click Apply Hide/Isolate to the view.

Exercise 4-11
Segmented Views

Drawing Name: **segmented views.rvt**
Estimated Time to Completion: 10 Minutes

Scope
Modify a section view into segments

Solution

1.

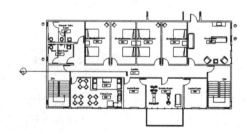

Activate the **A101- Segmented Elevation** sheet.

There are two views on the sheet. One is the floor plan and the other is a section view defined in the floor plan.

2.

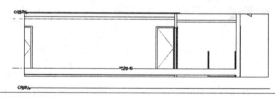

Activate the **Main Floor** floor plan.

This is the top view on the sheet.

3. Select the section line.

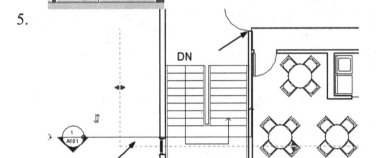

4. Select **Split Segment** on the ribbon.

5. Place a cut to the right of the stairs.
Drag the section line segment below the stairs.

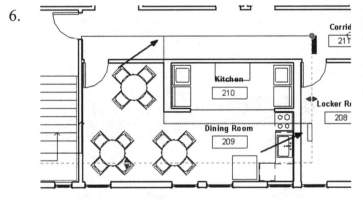

6. Place a cut to the left of the kitchen area.

Drag the section line below the oven in the kitchen.

7. Right click and cancel out the command.
The new section line is now segmented.

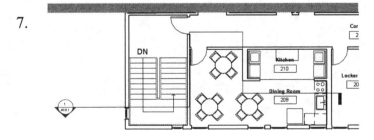

8. 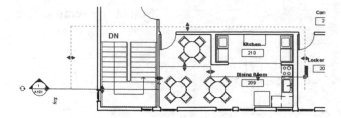 Reverse the direction of the section line so the arrow is pointed up.

9. Adjust the segments of the section line so the section lines are as shown.

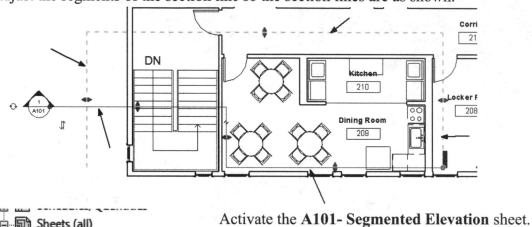

10. 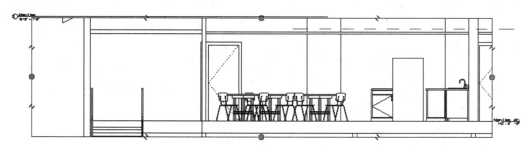 Activate the **A101- Segmented Elevation** sheet.

Note how the section view has updated.

11. Close without saving.

Exercise 4-12

Depth Cueing

Drawing Name: **depth cueing.rvt**
Estimated Time to Completion: 20 Minutes

Scope
Using different graphic display options.

Solution

1.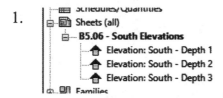

   Activate the **South Elevations sheet.**

   Note there are three identical views showing the South Elevation.

   Depth Cueing can only be applied in elevation views.

2.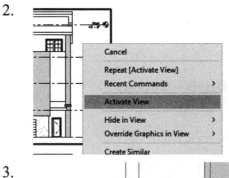

   Select the top view.
   Right click and select **Activate View**.

3.

   Select **Graphic Display Options**.

4.

Change the Model Display Style to **Realistic**.
Enable **Show Edges**.
Enable **Smooth lines with anti-aliasing**.
Set the Silhouettes to **Thin Lines**.
Press **Apply** to see the changes.

5.

Under Shadows:
Enable **Cast Shadows**.
Enable **Show Ambient Shadows**.
Press **Apply** to see the changes.
Press **OK**.

6.

Right click and **Deactivate** the view.

7.

Select the middle view.
Right click and select **Activate View**.

8.

Select **Graphic Display Options**.

9.

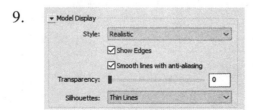

Change the Model Display Style to **Realistic**.
Enable **Show Edges**.
Enable **Smooth lines with anti-aliasing**.
Set the Silhouettes to **Thin Lines**.
Press **Apply** to see the changes.

10.

Under Shadows:
Enable **Cast Shadows**.
Enable **Show Ambient Shadows**.
Press **Apply** to see the changes.

11.

Under Depth Cueing:
Enable **Show Depth**.
Set Near to **50**.
Press **Apply** to see the changes.
Press **OK**.

12.

Right click and **Deactivate** the view.

13.

Compare how the views look different.

14.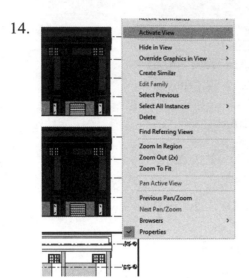

Select the bottom view.
Right click and select **Activate View**.

15.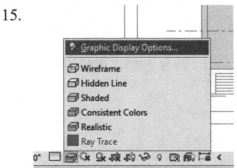

Select **Graphic Display Options**.

16.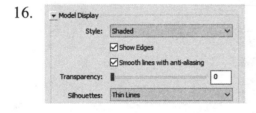

Change the Model Display Style to **Shaded**.
Enable **Show Edges**.
Enable **Smooth lines with anti-aliasing**.
Set the Silhouettes to **Thin Lines**.
Press **Apply** to see the changes.

17.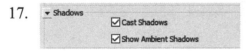

Under Shadows:
Enable **Cast Shadows**.
Enable **Show Ambient Shadows**.
Press **Apply** to see the changes.

18.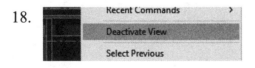

Right click and **Deactivate** the view.

19. Compare the three views.

20. Close without saving.

Exercise 4-13
Rotate a Cropped View

Drawing Name: **rotate_view.rvt**
Estimated Time to Completion: 10 Minutes

Scope
To rotate a plan view

Solution

1.

   Floor Plans
    Ground Floor
    Lower Roof
    Main Floor
     Main Floor - Kitchen
    Main Roof
    Site
    T. O. Footing
    T. O. Parapet

Open the **Main Floor – Kitchen** floor plan view.

2.

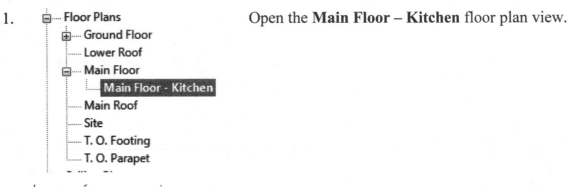

Verify the Crop Region is enabled as **Visible**.

3.

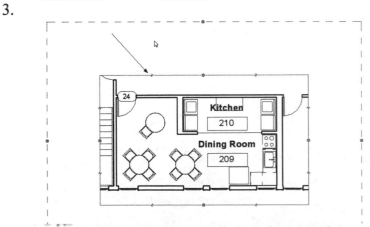

Select the crop region outline.

4. Select the **Rotate** tool on the Modify panel of the ribbon.

5.

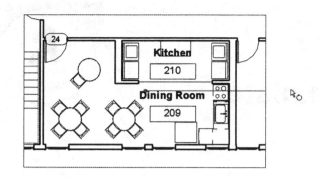

Set the starting point at 0 degrees.

6.

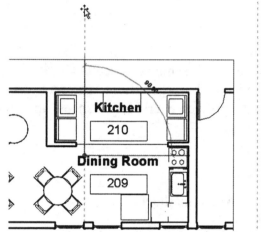

Set the ending point at 90 degrees.

7.

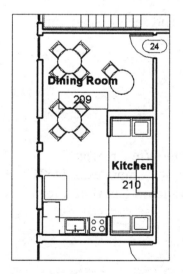

Adjust the crop region's size to show the kitchen and dining room area.

Move the room tags as needed.

Did you notice that the room tags rotated with the view?

8. Close without saving.

Exercise 4-14

Apply a View Template to a Sheet

Drawing Name: **view templates2.rvt**
Estimated Time to Completion: 10 Minutes

Scope
Use a view template to set all the views on a sheet to the same view scale.
View templates are used to standardize project views. In large offices, different people are working on the same project. By using a view project, everybody's sheets and views will be consistent.

Solution

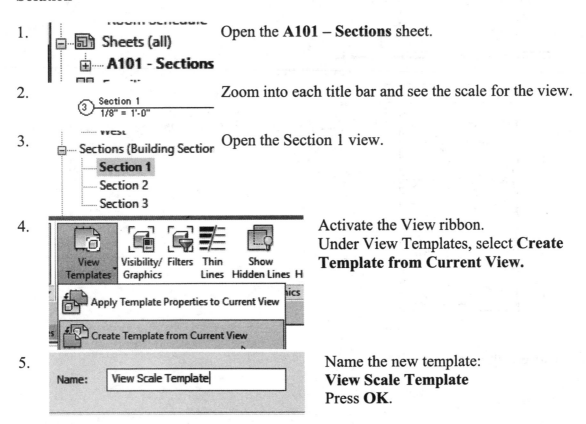

1. Open the **A101 – Sections** sheet.

2. Zoom into each title bar and see the scale for the view.

3. Open the Section 1 view.

4. Activate the View ribbon.
   Under View Templates, select **Create Template from Current View.**

5. Name the new template:
   View Scale Template
   Press **OK**.

6.

Parameter	Value	Include
View Scale	1/8" = 1'-0"	☑
Scale Value 1:	96	
Display Model	Normal	☐
Detail Level	Medium	☐
Parts Visibility	Show Original	☐
V/G Overrides Model	Edit...	☐
V/G Overrides Annotati	Edit...	☐

Uncheck all the boxes EXCEPT View Scale.
Press **OK**.

7.

Open the **A101 – Sections** sheet.

8.

Highlight the **A101** sheet in the browser.
Right click and select **Apply View Template to All Views.**

9.

Names:
Architectural Elevation
Architectural Section
Site Section
Structural Framing Elevation
Structural Section
View Scale Template

Select the **View Scale Template**.
Press **OK**.

10. All the views on the sheet adjusted. Re-position the views.

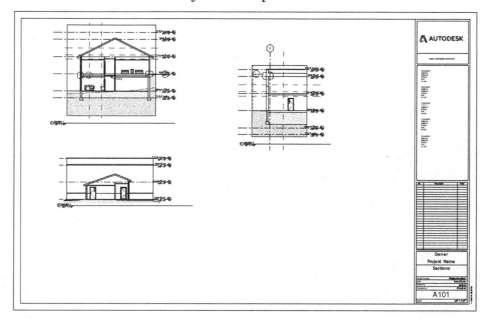

11. 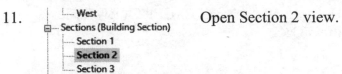 Open Section 2 view.

12.

Graphics	
View Scale	1/8" = 1'-0"
Scale Value 1:	96
Display Model	Normal

Note that the view scale has been modified for the view.

13. Close without saving.

Certified User Practice Exam

1. Straight grid lines are visible in the following view types:

   A. ELEVATION
   B. PLAN
   C. 3D
   D. SECTION
   E. DETAIL

2. Which two shortcut keys launch the Visibility/Graphics dialog?

   A. VG
   B. VV
   C. VE
   D. VW
   E. F5

3. **True or False**: Objects hidden using the Temporary Hide/Isolate tool are not visible, but they are still printed.

4. Select the THREE options for Detail Level for a view:

   A. COARSE
   B. SHADED
   C. FINE
   D. HIDDEN
   E. WIREFRAME
   F. MEDIUM

5. **True or False**: If you delete a view, the annotations placed in the view are also deleted.

6. **True or False**: A camera cannot be placed in an elevation view.

7. To change the graphic appearance of your model from Hidden Line to Realistic, you:

   A. Modify the Rendering Settings in the View Control Bar
   B. Edit Visibility/Graphics Overrides
   C. Change graphic display options in View Properties
   D. Click Visual Styles on the View Control Bar

8. A story level is the color:

   A. Yellow
   B. Blue
   C. Black
   D. Green

9. If you create a level using the COPY or ARRAY tool, what type of level is created?

   A. Story
   B. Non-Story or Reference
   C. Elevation
   D. Plan

10. This type of level does not have a PLAN view associated to it:

   A. Story
   B. Non-Story or Reference
   C. Elevation
   D. Plan

11. In which view type can you place a level?

   A. PLAN
   B. LEGEND
   C. 3D
   D. CONSTRUCTION
   E. ELEVATION

12. To display or open a view: (Select all valid answers)

   A. Double click the view name in the project browser
   B. Double click the elevation arrowhead, the section head or the callout head
   C. Right click the elevation arrowhead, the section head or the callout head, and select Go to View
   D. Select the Surf tool from the ribbon

Answers:
1) A, B & D; 2) A & B; 3) True; 4) A, C & F; 5) True; 6) False; 7) D; 8) B; 9) B; 10) B; 11) E; 12) A, B, and C

Certified Professional Practice Exam

1. Scope boxes control the visibility of:

    A. Elements
    B. Plumbing Fixtures
    C. Object Styles
    D. Grid lines and levels

2. If you rotate a cropped view: (Select the answer which is TRUE)

    A. Any tags will automatically rotate with the view.
    B. The crop region will automatically resize.
    C. The Project North will change.
    D. Any hidden elements will become visible.

3. Depth Cueing applies to all the elements listed EXCEPT: (Select two answers)

    A. Model elements, like walls, doors, and floors.
    B. Shadows and Sketchy Lines
    C. Annotations, like dimensions and tags
    D. Linework and line weight

4. A _____ is a collection of view properties, such as view scale, discipline, detail level and visibility settings.

    A. View template
    B. Graphics display
    C. Crop Region
    D. Sheet

5. _____ provide(s) a way to override the graphic display and control of the visibility of elements that share common properties in a view.

    A. Hide/Isolate
    B. Hide Category
    C. Hide Element
    D. Filters

6. Select the Type properties for a Level. (Select all correct answers)

    A. Elevation Base
    B. Line Weight
    C. Elevation
    D. Name

7. Select the ONE item you cannot create a view type:

    A. Floor Plans
    B. Ceiling Plans
    C. Area Plans
    D. 3D Views

8. A _____ displays the view number and sheet number (if the view is included on a sheet) in the corresponding view.

    A. Title bar
    B. View Name
    C. Sheet
    D. View Reference

Answers:
1) D; 2) A; 3) C & D; 4) A; 5) D; 6) A& B – C & D are instance properties; 7) C; 8) D

Dimensions and Constraints

This lesson addresses the following exam questions:

- Constraints
- Temporary and Permanent Dimensions
- Multi-segmented Dimensions

Dimensions are system families. They are in the Annotation category. They have type and instance properties.

Revit has three types of dimensions: listening, temporary and permanent.

A listening dimension is the dimension that is displayed as you are drawing, modifying, or moving an element.

A temporary dimension is displayed when an element is placed or selected. In order to modify a temporary dimension, you must select the element.

A permanent dimension is placed using the Dimension tool. In order to modify a permanent dimension, you must move the element to a new position. The permanent dimension will automatically update. To reposition an element, you can modify the temporary dimension or move the element using listening dimensions.

When you enter dimension values using feet and inches, you do not have to enter the units. You can separate the feet and inches values with a space and Revit will fill in the units. If a single unit is entered, for example '10', Revit assumes that value is 10 feet, not 10 inches.

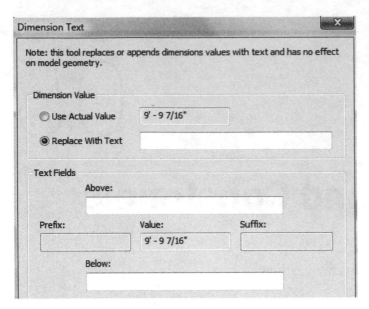

To override a dimension value, select the permanent dimension and enable Replace with Text and put in the desired text.

You can also add additional notes using the Text Fields for Above, Prefix, Suffix, or Below.

Exercise 5-1

Placing Permanent Dimensions

Drawing Name: **i_dimensions.rvt**
Estimated Time to Completion: 20 Minutes

Scope
Placing dimensions

Solution

1. Floor Plans
    — Ground Floor
    — **Ground Floor Admin Wing**
    — Lower Roof
    — Main Floor
    — Main Floor Admin Wing

    Activate the **Ground Floor Admin Wing** floor plan.

2. Activate the Modify ribbon.
    Select the **Match Properties** tool from the Clipboard panel.

3. Select the wall indicated as the source object.

4. 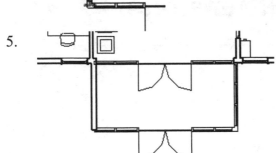 Select the wall indicated as the target object.

5. Note that the curtain wall changed to an Exterior-Siding wall.
Cancel out of the Match Properties command.

6. Activate the Annotate ribbon.
Select the **Aligned Dimension** tool.

Aligned

7. On the Options bar, set the dimensions to select the **Wall faces**.

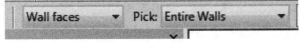

8. On the Options bar, set the dimensions to **Pick Entire Walls**.

9. Pick the wall indicated. Move the mouse above the selected wall to place the dimension.

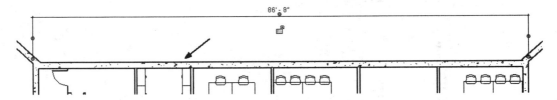

10. Note the entire wall is selected and the dimension is located at the faces of the walls.
Cancel or escape to end the Dimension command.

11. Select the dimension.
Note that there are several grips available. The grips are the small blue bubbles.

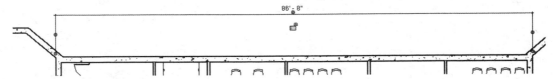

12. 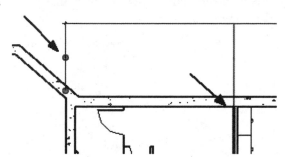 Select the second grip indicated on the left.

Move the witness line to the wall face indicated by the arrow in the center of the image.

Note that the witness line automatically snaps to the wall face.

13. 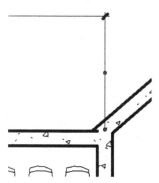 With the dimension selected:

On the Options bar, change the Prefer to **Wall centerlines**.

14. 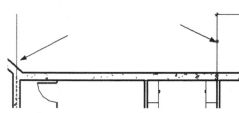 Select the dimension and activate the grip indicated.

Drag the witness line to the center of the left wall.

15.  Left click once on the grip on the right side of the dimension to shift the witness line to the wall centerline.

16. Note how the dimension value updates.

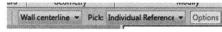

85' - 8"

17. Select the **Aligned** tool.

Aligned

18. Set Pick to Individual References on the options bar.

Wall centerline ▾ Pick: Individual Reference ▾ [Options]

19.

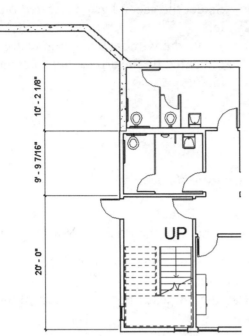

Select the centerlines of the walls indicated.

20.

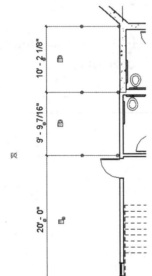

Select the dimension so it highlights.

Select the two locks on the top dimensions to switch the permanent dimensions to *locked* dimensions. This means these distances will not be changed.

You should see two of the padlocks as closed and one as open.

21.

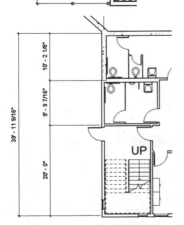

Place an overall dimension using the wall centerlines.

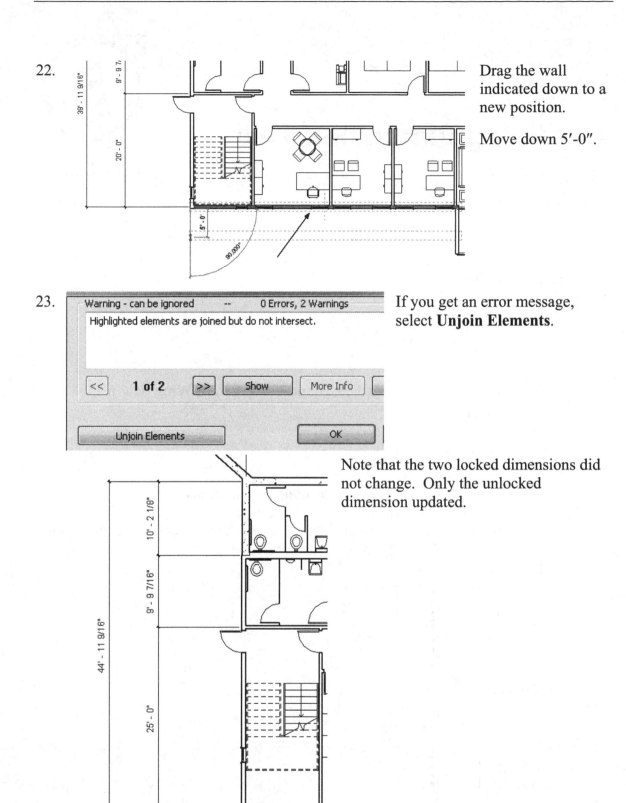

22.   Drag the wall indicated down to a new position.

Move down 5'-0".

23.   If you get an error message, select **Unjoin Elements**.

Note that the two locked dimensions did not change. Only the unlocked dimension updated.

24. Close without saving.

Exercise 5-2
Modifying Dimension Text

Drawing Name: **dimtext.rvt**
Estimated Time to Completion: 15 Minutes

Scope
Replace Dimension Text
Restore Dimension Text
Modify Dimension Text

Solution

1. Activate the **Ground Floor Admin Wing** floor plan.

   Floor Plans
   Ground Floor
   Ground Floor Admin Wing
   Lower Roof
   Main Floor
   Main Floor Admin Wing
   Main Roof

2.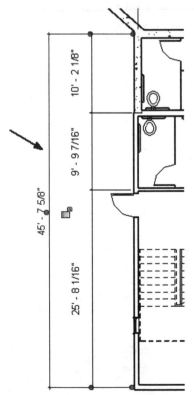

   Select the dimension indicated.

   Double left click on the dimension text to bring up the dimension text dialog.

3.

Dimension Value

○ Use Actual Value   45' - 7 5/8"

◉ Replace With Text   45' 8"

Enable **Replace with Text**.
Enter **45′ 8″**.
Press **OK**.

4.

Invalid Dimension Value   ✕

Specify descriptive text for a dimension
segment instead of a numeric value.

To change the dimension value for a length or angle of a
segment, select the element the dimension refers to, and click
the value to edit it.

You will get an error message stating that
you cannot change the numeric value
without moving the element to correspond
with that value.

Press **Close**

5.

Dimension Value

○ Use Actual Value   45' - 7 5/8"

◉ Replace With Text   OVERALL

Enable **Replace with Text**.
Enter **OVERALL**.
Press **OK**.

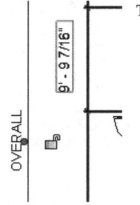

The dimension text updates.

6.

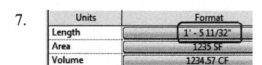

Project
rs Units

Activate the **Manage** ribbon.
Select **Project Units** on the Settings panel.

7.

Units	Format
Length	1' - 5 11/32"
Area	1235 SF
Volume	1234.57 CF

Select the **Length** button under the Format
column.

8.

Units:

Rounding:

To the nearest 1/2"

Set the Rounding **To the nearest ½″**.
Press **OK** until all dialogs are closed.

Note that the dimensions update to the new rounding.

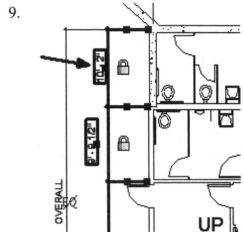

9.   Double left click on the 10′ 2″ dimension to be edited.

10. In the Below field, type:
WOMEN'S LAVATORY.
Press **OK**.

11. Note how the dimension updates.

12.   Double left click on the dimension with **???**.

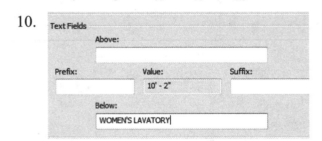

13. Enable **Use Actual Value**.
Press **OK**.

14. Close without saving.

Exercise 5-3
Converting Temporary Dimensions to Permanent Dimensions

Drawing Name: **i_dimensions.rvt**
Estimated Time to Completion: 10 Minutes

Scope
Place dimensions using different options.
Convert temporary dimension to permanent

Solution

1.
   Floor Plans
       Ground Floor
       Ground Floor Admin Wing
       Lower Roof
       Main Floor
       Main Floor Admin Wing
       Main Roof
       Site

   Activate the **Ground Floor Admin Wing** floor plan.

2. Select the wall indicated.

3. Two dimensions will appear. There is a small dimension icon visible.
   This icon converts a temporary dimension to a permanent dimension.
   Left click on this icon.

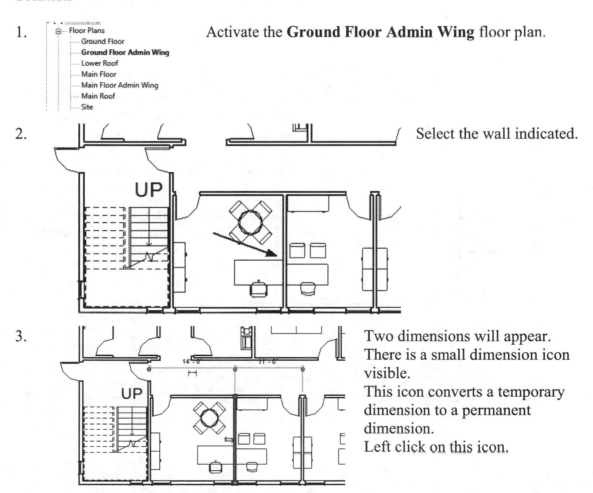

4.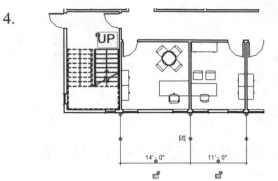

The dimensions are now converted to permanent dimensions.
Drag the dimensions below the view.

5.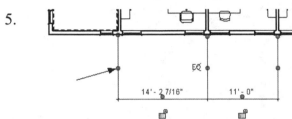

Click on the dimensions so they highlight.
Click on the witness line grip indicated.

Notice how each time the grip is clicked on, the witness line shifts from the wall centerline to the face of the wall.

6.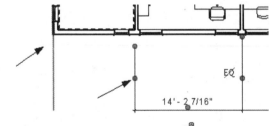

Select the grip at the endpoint of the witness line. Drag the endpoint to increase the gap between the wall and the witness line.
Release the grip.

7.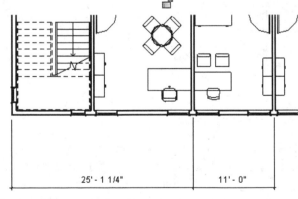

Select the witness line grip indicated.
Drag the witness line to the outer wall.
Note that if the witness line is set to the outer wall face it maintains that orientation.
Release.

The dimension updates.
Note the new dimension value.

8. Close without saving.

Exercise 5-4
Applying Constraints

Drawing Name: **i_constraints.rvt**
Estimated Time to Completion: 20 Minutes

Scope
Using the Align tool to constrain elements

Solution

1. Floor Plans
    Ground Floor
    Lower Roof
    Main Floor
    Main Roof

    Activate the **Main Floor** floor plan.

2.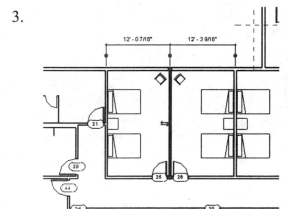

    Select the interior wall located between Door 25 and Door 26 as indicated.

3. The associated temporary dimensions will become visible.
    Left click on the permanent dimension toggle to convert the dimensions to permanent dimensions.

5-13

4. Left click anywhere in the display window.
In the Properties pane:

Scroll down to the **Underlay** parameter.
Set the Underlay to **Ground Floor**.

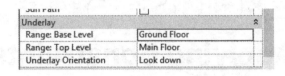

Underlay	
Range: Base Level	Ground Floor
Range: Top Level	Main Floor
Underlay Orientation	Look down

This will make the ground floor visible in the graphics window. Elements on the ground floor will be displayed in a lighter shade.

5. Note in the upper left corner the interior walls are not aligned.

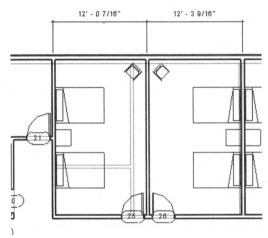

6. Activate the **Modify** ribbon.
Select the **Align** tool from the Modify panel.

7. Set the preference to **Wall faces** on the Options bar.

8. Select the right side of the ground floor interior wall as the source for the alignment.

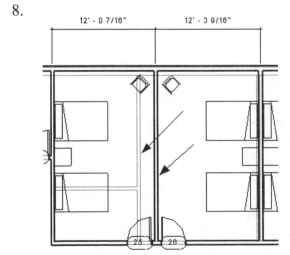

9.

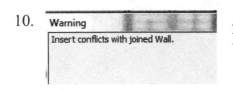

Select the right side of the main floor interior wall as the target for the alignment (this is the wall that will be shifted).

10.

Warning

Insert conflicts with joined Wall.

A warning dialog will appear due to the door location. Ignore the warning. ***Don't move the door or wall.***

11.

Lock the alignment in place so that the walls remain constrained together.

12. Right click and select Cancel to exit the Align mode.

13.

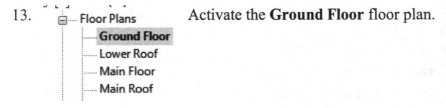

Activate the **Ground Floor** floor plan.

14.

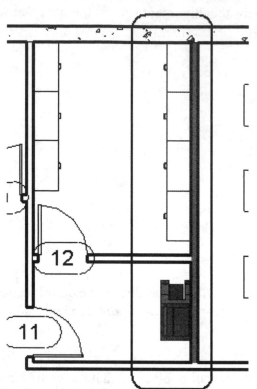

Select the wall, the cabinets, and copier next to the wall.

15. Select the **Move** tool on the Modify ribbon.

16.

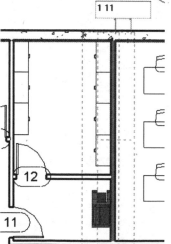

1 11

Select a base point.

Move the selected elements to the right **1′ 11″**.

Left click in the window to release the selected elements.

17. Floor Plans
   Ground Floor
   Lower Roof
   Main Floor
   Main Roof
   Site
   T. O. Footing
   T. O. Parapet

Activate the **Main Floor** floor plan.

18.

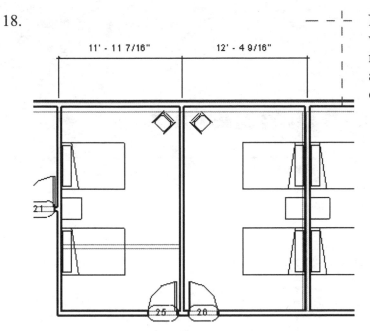

Note that the main floor interior wall has also shifted (the walls remained aligned because they are locked together) and the dimension has updated.

19. Close without saving.

Revit uses three types of constraints:

- Explicit
- Loose
- Implied

An explicit constraint determines that defined relationships are always maintained. An example of an explicit constraint would be when two elements are aligned and then locked or when a dimension is locked.

A loose constraint is a dimension or alignment which is not locked. They are maintained unless a conflict occurs.

Implied constraints are also maintained unless a conflict occurs, such as when a wall is attached to a roof or two walls are joined at a corner.

An Equality constraint is considered an explicit constraint.

You can display constraints in a view by selecting Reveal Constraints on the View Control bar.

Exercise 5-5
Multi-Segmented Dimensions

Drawing Name: **multisegment.rvt**
Estimated Time to Completion: 10 Minutes

Scope
Add and delete witness lines to a multi-segment dimension

Solution

1. Activate the **Main Floor** floor plan.

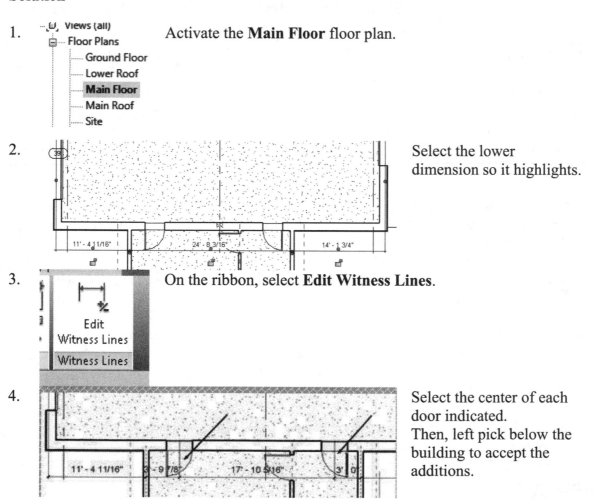

2. Select the lower dimension so it highlights.

3. On the ribbon, select **Edit Witness Lines**.

4. Select the center of each door indicated.
Then, left pick below the building to accept the additions.

5.
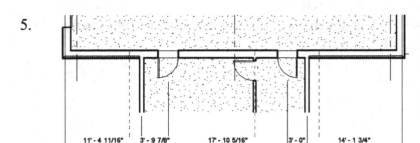
Drag the dimensions down so you can see them easily.

6.
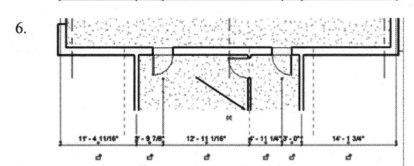
Select the dimension.
Select **Edit Witness Lines**.
Add the witness line at the wall indicated.

7.
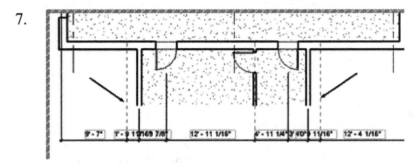
Select the dimension.
Select **Edit Witness Lines**.
Add the witness line at the two reference planes indicated.

8.
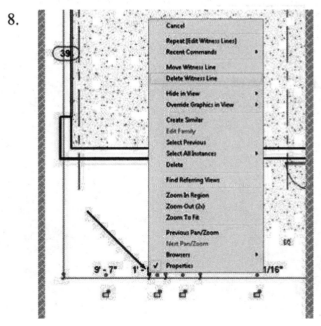
Select the dimension.
Select the grip on the witness line for the left reference plane.
Right click and select **Delete Witness Line**.

9.

The dimension should update with the witness line removed.

11' - 4 11/16"   3' - 9 7/8"   12' -

10.

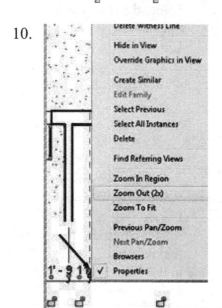

Repeat to delete the witness line on the right reference plane.

11.

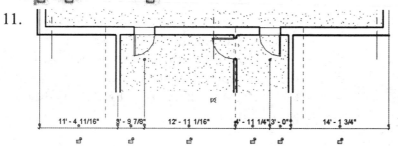

The dimension should update.

11' - 4 11/16"   3' - 9 7/8"   12' - 11 1/16"   4' - 11 1/4" 3' - 0"   14' - 1 3/4"

12. Close without saving.

Exercise 5-6
Equality Formula

Drawing Name: **equality_formula.rvt**
Estimated Time to Completion: 10 Minutes

Scope
Using an Equality Formula

Solution

1. Floor Plans
   — Ground Floor
   — Lower Roof
   Main Floor
   — Main Roof

   Activate the **Main Floor** floor plan.

2. Select the EQ dimensions.
   Right click and disable EQ display.
   Note the value is 16' 4''.

   Cancel
   Repeat Last Command
   Edit Witness Lines
   EQ Display
   Flip Dimension Direction
   Reset Dimension Text Position
   Hide in View

3. Properties

   Linear Dimension Style
   Linear - 3/32" Arial

   Dimensions (1)
   Graphics
   Leader
   Baseline Offset... 0"
   Other
   Equality Display Value
   Value
   Equality Text
   Equality Formula

   With the dimension still selected, go to the Properties pane.
   Under Equality Display:
   Select **Equality Formula**.

4. Linear Dimension Style
   Linear - 3/32" Arial

   Dimensions (1) Edit Type
   Graphics
   Leader
   Baseline Offset...
   Other
   Equality Display Equality Formula

   Select **Edit Type**.

5. Other
   Equality Text EQ
   Equality Formula Total Length
   Equality Witness Display Tick and Line

   Scroll down. Select the button next to Equality Formula.

6.

	Parameter Name	Spaces	Pref
1	Number of Segments	1	
2	Length of Segment	1	
3	Total Length	0	

Add Number of Segments and Length of Segment to Label Parameters.
Organize so that Number of Segments is on Row 1; Length of Segment is on Row 2, and Total Length is on Row 3.

7.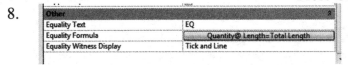

	Parameter Name	Spac	Prefix	
1	Number of Segments	1		@
2	Length of Segment	1		=
3	Total Length	0		

Add a Suffix for Number of Segments - @.
Add a Suffix for Length of Segment - =.
Place a space in each Prefix field.

This builds an equation that reads

<Number of Segments> @ <Length of Segment> = <Total Length>

Press **OK**.

8.

Other	
Equality Text	EQ
Equality Formula	Quantity@ Length= Total Length
Equality Witness Display	Tick and Line

The Equality Formula is updated.
Press **OK**.

9.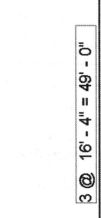

Left click in the graphics window to release the selection and update the dimension.

If you don't see spaces, go back and add spaces to the prefix fields.

10. Close without saving.

Exercise 5-7
Alternate Dimensions

Drawing Name: **Alternate Dimensions.rvt**
Estimated Time to Completion: 10 Minutes

Scope
Apply Alternate Dimensions

Solution

1.
   - Views (all)
     - Floor Plans
       - Ground Floor
       - Lower Roof
       - **Main Floor**
       - Main Roof
       - Site

   Activate the **Main Floor** floor plan.

2.

   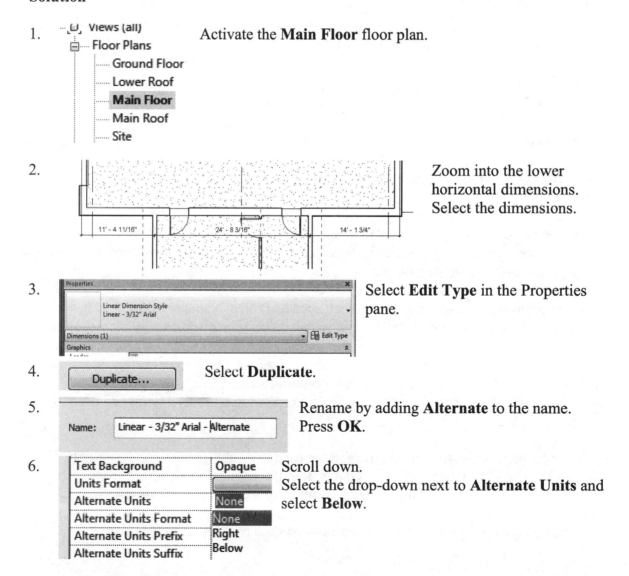

   Zoom into the lower horizontal dimensions. Select the dimensions.

   11' - 4 11/16"   24' - 8 3/16"   14' - 1 3/4"

3.

   Properties

   Linear Dimension Style
   Linear - 3/32" Arial

   Dimensions (1)   Edit Type
   Graphics

   Select **Edit Type** in the Properties pane.

4.

   Duplicate...

   Select **Duplicate**.

5.

   Name: Linear - 3/32" Arial - Alternate

   Rename by adding **Alternate** to the name. Press **OK**.

6.

Text Background	Opaque
Units Format	
Alternate Units	None
Alternate Units Format	None
Alternate Units Prefix	Right
Alternate Units Suffix	Below

   Scroll down.
   Select the drop-down next to **Alternate Units** and select **Below**.

7.

Text Background	Opaque
Units Format	1' - 5 11/3
Alternate Units	Below
Alternate Units Format	1235
Alternate Units Prefix	
Alternate Units Suffix	mm
Show Opening Height	☐
Suppress Spaces	☐

Type **mm** in the Alternate Units Suffix field.

Press **OK**.

8.

11' - 4 11/16"

3472mm

The dimensions now display with both Imperial and millimeter units.

9.   Close without saving.

Global Parameters

Revit uses four different types of parameters:

- Family
- Project
- Shared
- Global

Users should be familiar with all the different types of parameters and how they are applied.

Family Parameters are used in loadable families, such as doors and windows. They normally don't appear in schedules or tags. They are used to control size and materials. They can be linked in nested families.

Project parameters are specific to a project, such as Project Name or address. They may appear in schedules or title blocks but not in tags. They can also be used to categorize views.

Shared parameters is a *.txt file placed in a central location, like a server. It lists parameters which can be applied to any project or used in a family. It may be used in schedules or tags.

Global parameters are project specific and are used to establish rules for dimensions – such as the distance between a door and a wall or the width of a corridor or room size. They can also be used in formulas. Global parameters can be driving or reporting. A reporting parameter is controlled by a formula. Length, radius, angle or arc length can all be used by reporting parameters. Area cannot be used by a reporting parameter.

Exercise 5-8
Global Parameters

Drawing Name: **global parameters.rvt**
Estimated Time to Completion: 10 Minutes

Scope
Using global parameters in a project.

Solution

1.
    Floor Plans
    Level 1
    Level 2
    Level 3
    Site

    Activate the **Level 2** floor plan.

2. Activate the Manage ribbon.
    Select **Global Parameters**.

3. Select **New Global Parameter**.

4.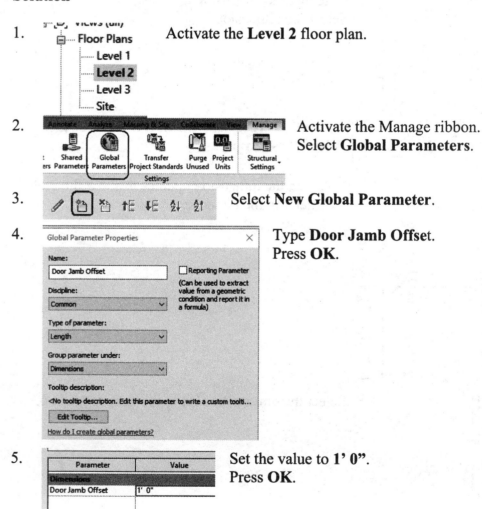

    Global Parameter Properties

    Name:
    Door Jamb Offset

    ☐ Reporting Parameter
    (Can be used to extract value from a geometric condition and report it in a formula)

    Discipline:
    Common

    Type of parameter:
    Length

    Group parameter under:
    Dimensions

    Tooltip description:
    <No tooltip description. Edit this parameter to write a custom toolti...

    Edit Tooltip...

    How do I create global parameters?

    Type **Door Jamb Offse**t.
    Press **OK**.

5. | Parameter | Value |
    |-----------|-------|
    | **Dimensions** | |
    | Door Jamb Offset | 1' 0" |

    Set the value to **1' 0"**.
    Press **OK**.

6. 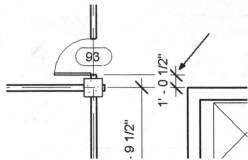 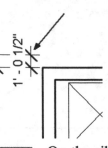 Select the dimension indicating the door jamb offset for the door with the tag labeled 93.

7. 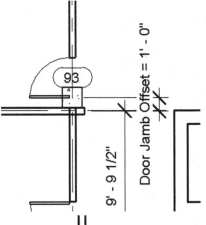 On the ribbon, select **Door Jamb Offset** under Label.

8. 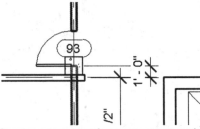 Note that the dimension updates.
Select the dimension.

9. In the Properties panel: Enable **Show Label in View.** Press **Apply**.

Other	
Label	Door Jamb Offset
Show Label in View	☑

10. Note that the dimension updates to show the global parameter in the dimension.

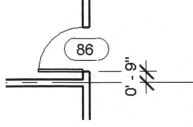

11. Select the dimension next to the door tagged 86.

12.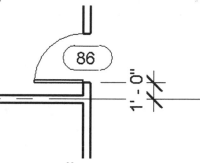

In the Properties panel:
Apply the Door Jamb Offset label.

13.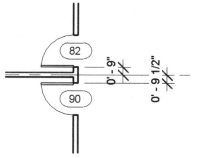

Note that the dimension updates.

14.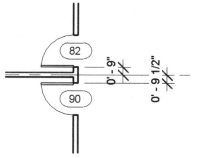

Zoom into the doors tagged 82 and 90.

15.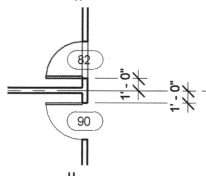

Modify the dimensions using the global parameter called Door Jamb Offset.

16.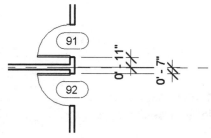

Zoom into the doors tagged 82 and 90.

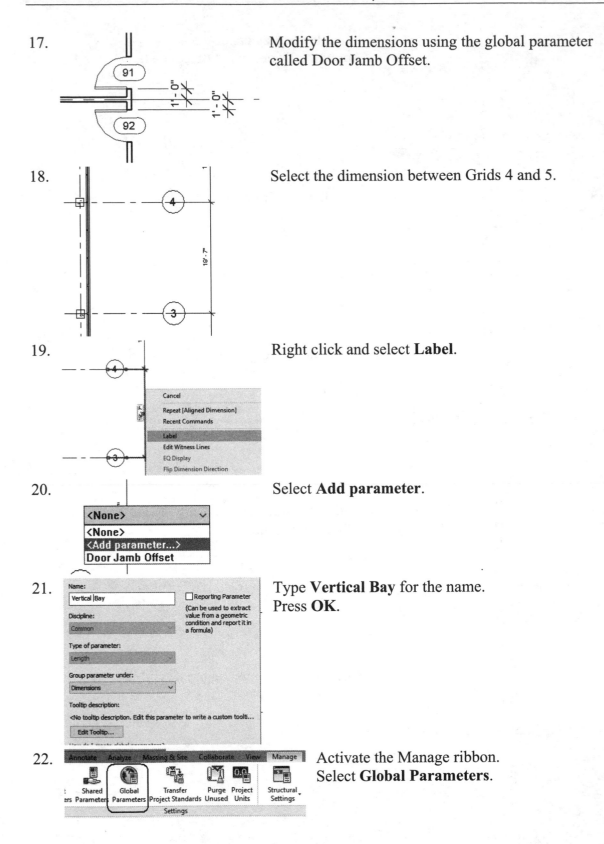

17.   Modify the dimensions using the global parameter called Door Jamb Offset.

18.   Select the dimension between Grids 4 and 5.

19.   Right click and select **Label**.

20.   Select **Add parameter**.

21.   Type **Vertical Bay** for the name.
       Press **OK**.

22.   Activate the Manage ribbon.
       Select **Global Parameters**.

23.

Parameter	Value
Dimensions	
Door Jamb Offset	1' 0"
Vertical Bay	18' 2"

Set the value for the Vertical Bay to **18' 2"**.

24.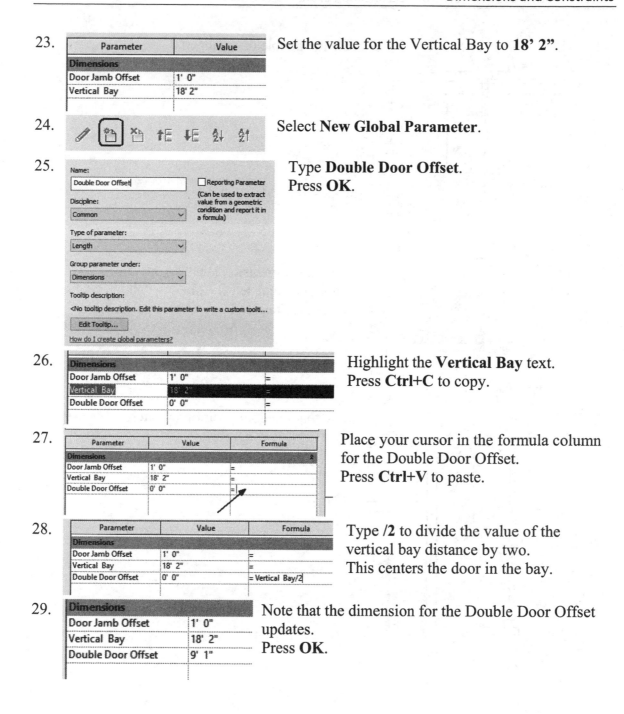

Select **New Global Parameter**.

25.

Name:

Double Door Offset

☐ Reporting Parameter

(Can be used to extract value from a geometric condition and report it in a formula)

Discipline:

Common

Type of parameter:

Length

Group parameter under:

Dimensions

Tooltip description:

<No tooltip description. Edit this parameter to write a custom toolti...

Edit Tooltip...

How do I create global parameters?

Type **Double Door Offset**.
Press **OK**.

26.

Dimensions		
Door Jamb Offset	1' 0"	=
Vertical Bay	18' 2"	=
Double Door Offset	0' 0"	=

Highlight the **Vertical Bay** text.
Press **Ctrl+C** to copy.

27.

Parameter	Value	Formula
Dimensions		
Door Jamb Offset	1' 0"	=
Vertical Bay	18' 2"	=
Double Door Offset	0' 0"	=

Place your cursor in the formula column for the Double Door Offset.
Press **Ctrl+V** to paste.

28.

Parameter	Value	Formula
Dimensions		
Door Jamb Offset	1' 0"	=
Vertical Bay	18' 2"	=
Double Door Offset	0' 0"	= Vertical Bay/2

Type **/2** to divide the value of the vertical bay distance by two.
This centers the door in the bay.

29.

Dimensions	
Door Jamb Offset	1' 0"
Vertical Bay	18' 2"
Double Door Offset	9' 1"

Note that the dimension for the Double Door Offset updates.
Press **OK**.

30.

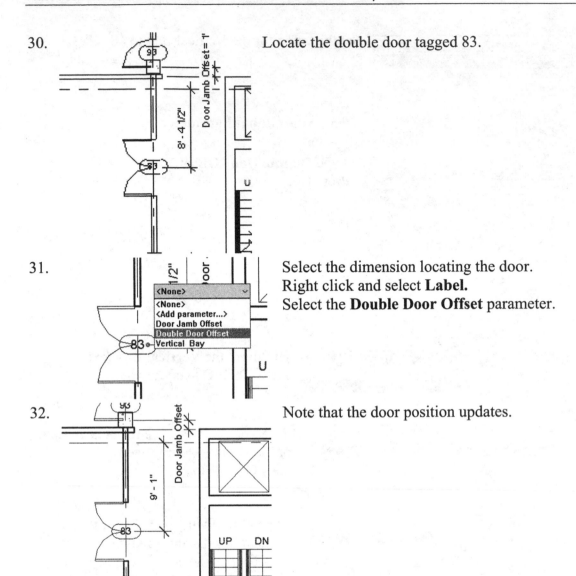

Locate the double door tagged 83.

31.

Select the dimension locating the door.
Right click and select **Label.**
Select the **Double Door Offset** parameter.

32.

Note that the door position updates.

33.

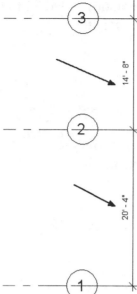 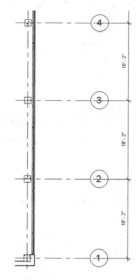

Assign the Vertical Bay parameter to the dimensions between grids 2 & 3 and grids 1 & 2.

34.

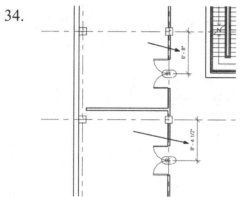

Apply the double door offset parameter to the doors labeled 84 and 85.

35.

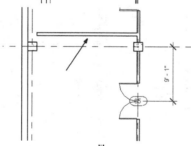

Use the ALIGN tool to move the wall above the door labeled 85 to center it on Grid 2.

36.

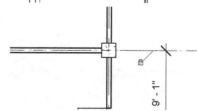

Lock the wall to the grid.

37. Use the ALIGN tool to position the wall above the door labeled 83 to Grid 4 and lock into position.

38. 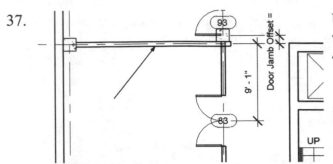 Activate the Manage ribbon.
    Select **Global Parameters**.

39.

Parameter	Value
Dimensions	
Door Jamb Offset	1' 0"
Vertical Bay	18' 6"
Double Door Offset	9' 1"

Change the value of the Vertical Bay to **18' 6"**.
Press **OK**.

40. Note that the dimensions update and the double doors remain centered in the bays. Close without saving.

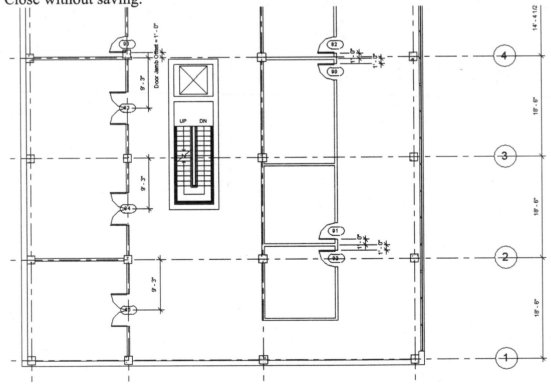

Exercise 5-9
Reveal Constraints

Drawing Name: **reveal_constraints.rvt**
Estimated Time to Completion: 5 Minutes

Scope
Using the Reveal Constraints tool.

Solution

1. Activate the **Main Floor** floor plan.

2. Select the **Reveal Constraints** tool on the View Control bar.

3. 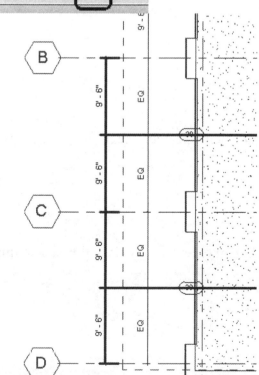 What dimension is displayed between Grids B & C?

   The answer is 9' 6".

4. Disable the Reveal Constraints tool on the View Control bar.
5. Close without saving.

Certified User Practice Exam

1. Constraints can be used to:

    A. Prevent other users from moving an element in a shared project.
    B. Lock two elements at a fixed distance.
    C. Keep two elements equally spaced.
    D. Alert the user when someone changes or moves an element.

2. To change a dimension's value:

    A. Use the Change Dimension tool.
    B. Click on Edit Witness Line.
    C. Select the element to be re-dimensioned.
    D. Click on the dimension.

3. **True or False**: You can override a dimension value with a new dimension value.

4. Select the TWO types of dimensions:

    A. PERMANENT
    B. TEMPORARY
    C. PERPETUAL
    D. INVISIBLE

5. To create a dimension using feet and inches, you can use this keystroke to separate the feet and inches values:

    A. COMMA
    B. SEMI-COLON
    C. SPACE
    D. COLON

6. A _____ allows you to locate elements equidistant from one another.

    A. multi-segmented dimension
    B. single-segmented dimension
    C. scope box
    D. reference plane

7. You are editing a dimension in Revit. You type 0 48. Revit will translate this to:

    A. 4' 0"
    B. 48' 0"
    C. 48"
    D. 4' 8"

8. You are placing a wall in Revit. You use the listening dimension to set the wall length. You enter '**5.5**'. Revit places the wall and it is _____ long.

    A. 5' 6"
    B. 5 1/2"
    C. 5' 0"
    D. 5' 5"

9. To change the position of an element, such as a door or wall:

    A. Place a permanent dimension, then modify the dimension
    B. Select the element, then modify the temporary dimension
    C. Select the element and move it
    D. Use the Measure tool

10. The three types of dimensions used in Revit are:

    A. LINEAR
    B. ANGULAR
    C. TEMPORARY
    D. PERMANENT
    E. CONTINUOUS
    F. LISTENING

Answers:
    1) B & C; 2) C; 3) False; 4) A & B; 5) C; 6) A; 7) A; 8) A; 9) B; 10) C, D, & F

Certified Professional Practice Exam

1. An equality formula can be applied to all these dimension types EXCEPT:

    A. Aligned
    B. Linear
    C. Arc
    D. Radius

2. _____ are non-view specific elements than can function independently of dimensions. These appear as blue dashed lines in all views in which their references are visible.

    A. Reference lines
    B. Reference planes
    C. Constraints
    D. Temporary dimensions

3. To display alternate dimensions, you need to modify the dimension's:

    A. Type Properties
    B. Instance Properties
    C. Dimension Constraints
    D. View Properties

4. Global parameters can be used for the following EXCEPT:

    A. Drive the value of a dimension or a constraint
    B. Associate to an element instance or type property to drive it's value
    C. Associate to an instance property of an element
    D. Drive the value of a variable used in a calculated value for a schedule

Answers:
    1) D; 2) C; 3) A; 4) D

Developing the Building Model

This lesson addresses the following User and Professional exam questions:

- Floors
- Ceilings
- Ceiling Lighting Fixtures
- Stairs
- Railings
- Roofs

Floors are level-based system families. A floor is added on the active level. The top of the floor is aligned to the level with its thickness projected downwards. You can also offset the floor so it is above or below a level using Element Properties.

Floors can be sloped. Floors can also be created by creating a mass element and selecting a face on the mass element.

Openings can be added to floors either by modifying the floor sketch or adding an opening. Railings can be added to either sloped or flat floors.

Ceilings are level-based system families. Ceilings can be controlled in a similar manner as floors using element properties.

Both ceilings and floors are Model elements.

Roofs are also level-based system families.

Stairs and railings are system families. They use profiles, which can be loaded. When you click the starting point of the stairs in plan view, the number of treads is calculated based on the distance between the floors and the maximum riser height. You can adjust the maximum riser height in the stairs element properties.

Railings consist of rails and balusters.

Stairs are automatically created assuming that they are going up from the level where they are placed. You can change the direction of the stairs (flip from going up to going down) by using the control arrow.

Exercise 6-1

Modifying a Floor Perimeter

Drawing Name: **i_floors.rvt**
Estimated Time to Completion: 15 Minutes

Scope
Modify a floor.
Determine the floor's perimeter.

Solution

1. Floor Plans
    ├── Ground Floor
    ├── Lower Roof
    ├── Main Floor
    ├── **Main Floor Admin Wing**
    ├── Main Roof
    └── Site

    Activate **the Main Floor Admin Wing** floor plan.

2. Window around the area indicated.

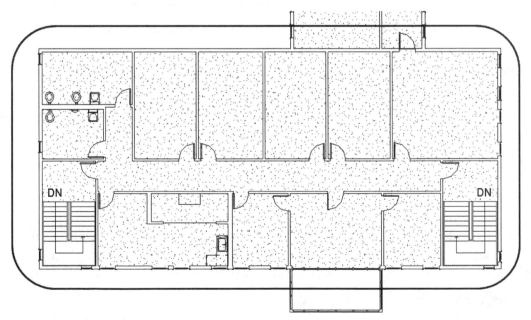

3. ☑ Press & Drag 🔽 :222

    Select the **Filter** tool located in the lower right corner of the window.

4.

Category:	Count:
☐ <Room Separation>	1
☐ Casework	5
☐ Curtain Panels	12
☐ Curtain Wall Grids	5
☐ Curtain Wall Mullions	16
☐ Doors	15
☑ **Floors**	1
☐ Lines (Lines)	16
☐ Plumbing Fixtures	7
☐ Railings	4
☐ Rooms	13
☐ Stairs	2
☐ Walls	27
☐ Windows	18
Total Selected Items:	1

Select **Check None** to disable all the checks.

Then check only the **Floors**.

Press **OK**.

5.

Dimensions	
Slope	
Perimeter	254' 8"
Area	3254.30 SF
Volume	3457.69 CF
Thickness	1' 0 3/4"

In the Properties pane:
Note that the floor has a perimeter of 254' 8".

6. 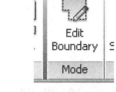 Select **Edit Boundary** under the Mode panel.

7. 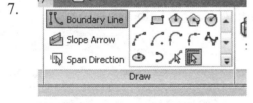 Select **Pick Walls** mode under the Draw panel.

8. Offset: 0' 0" ☐ Extend into wall (to core) Uncheck the Extend into wall (to core) option.

9. 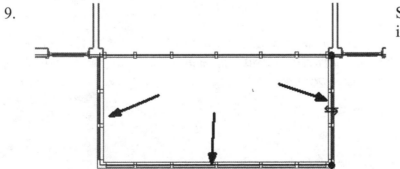 Select the three walls indicated.

10. Select the **Trim** tool on the Modify panel from the ribbon.

11.   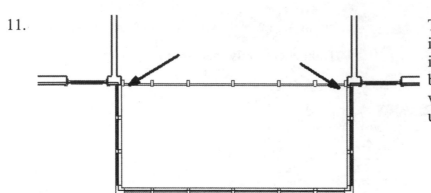   Trim the two corners indicated so that there is no wall in the section between the two vertical walls on the upper ends.

12.      Select the **Green Check** to **Finish Floor** from the ribbon.

13.   Revit

Would you like walls that go up to this floor's level to attach to its bottom?

Yes   No

Select **No**.

14.

Dimensions	
Slope	
Perimeter	270' 8"
Area	3390.71 SF
Volume	3602.63 CF
Thickness	1' 0 3/4"

Note that the floor has a perimeter of 270' 8". Press **OK** to close the dialog.

15.   Close without saving.

In the certification exam, you may be asked to modify a floor's sketch and then enter the resulting area, perimeter, or volume. Practice this exercise until you can get the correct perimeter value and feel comfortable using the tools.

Exercise 6-2

Modifying a Ceiling

Drawing Name: **i_ceilings.rvt**
Estimated Time to Completion: 15 Minutes

Scope
Modify a ceiling.
Rotate a ceiling grid.
Determine the ceiling's elevation.

Solution

1.   Ceiling Plans
          Ground Floor
          Ground Floor Admin Wing   Activate the **Ground Floor Admin Wing** ceiling plan.

2. Zoom into the area where the ceiling grid and lighting fixtures are placed.

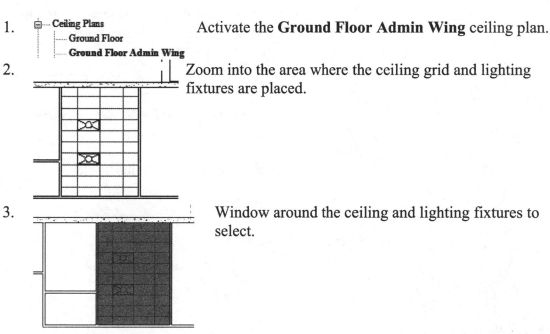

3. Window around the ceiling and lighting fixtures to select.

4. Select the **Filter** tool.

5. Verify that only the ceiling and lighting fixtures are selected.
Press **OK**.

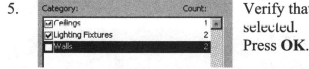

6.   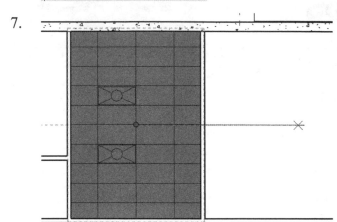   Select the **Rotate** tool from the Modify panel on the ribbon.

7.   Select a base point that is horizontal and to the right of the selected elements.

8.   Move the cursor above the selected elements. Enter **90** to ensure a 90-degree rotation.

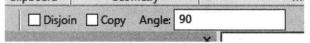

You can also enter 90 for the angle on the Options bar.

Press **ESC** to release the selection.

9.   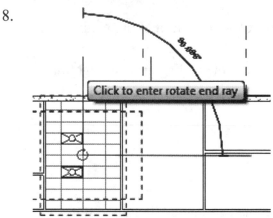   Select the ceiling so it highlights.

10.   Select **Edit Boundary** from the Mode panel.

11.  Select the **Align** tool from the Modify panel on the ribbon.

12. 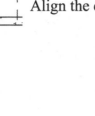 Align the ceiling boundary to the inner face of the walls.

13. Select the **Green Check** to **Finish Ceiling**.

14. 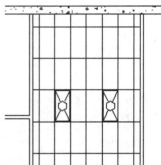 The grid updates.
Left click in the window to release the selection.

15. 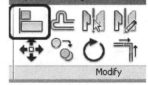 Select the **Align** tool from the Modify panel on the ribbon.

16. 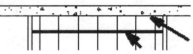 Select the inner face of the upper wall and align with the top horizontal gridline.

17. 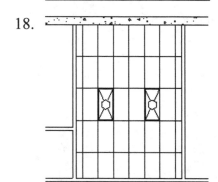 The ceiling grid updates so that all the grid tiles are all in the vertical direction.

18. Select the ceiling ONLY using Filter or the TAB functions.

19.

Constraints	
Level	Ground Floor
Height Offset From Level	8' 0"
Room Bounding	☑

In the Properties pane:
Note that the user can control the height offset from the level where the ceiling is placed.
Note the perimeter, area, and volume values.

20. Close without saving.

The certification exam will ask the user to either place or modify a ceiling and then identify one of the element parameters, such as perimeter, area, or volume. Users should practice this exercise until they are comfortable with the tools.

Exercise 6-3
Creating Stairs by Sketch

Drawing Name: **i_stairs.rvt**
Estimated Time to Completion: 20 Minutes

Scope
Place stairs using reference work planes.

Solution

1. Floor Plans
     — **Ground Floor**
     — Lower Roof
     — Main Floor
     — Main Roof
     — Site
     — T. O. Footing
     — T. O. Parapet

    Activate the **Ground Floor** floor plan.

2. Zoom into the lower left corner of the building.

3. Select the **Reference Plane** tool from the Work Plane panel on the Architecture ribbon.

    v Ref
       Plane

4. Offset: 2' 4"   Set the Offset to **2' 4"** on the Options bar.

5. 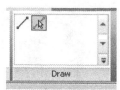 Select the **Pick Lines** tool from the Draw panel.

6. 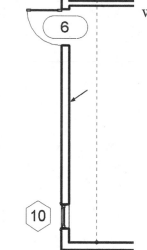 Place a vertical reference plane 2′ 4″ to the left of the wall where Door 4 is placed.

7. Place a vertical reference plane 2′ 4″ to the right of the wall where Door 6 is placed.

8. 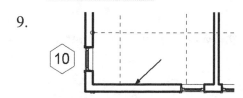 Set the Offset to **4′ 4″** on the Options bar.

9. Place a horizontal reference plane 4′ 4″ above the wall where Window 10 is located.

10. Set the Offset to **7′ 4″** on the Options bar.

11. 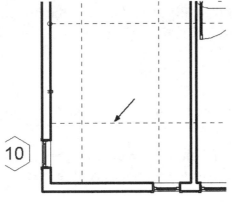 Place a horizontal reference plane 7' 4" above the first horizontal reference plane.

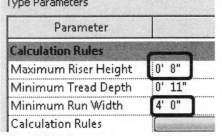 Four reference planes – two horizontal and two vertical should be placed in the room.

12. 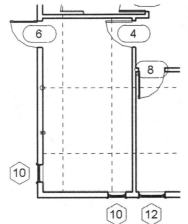 Activate the Architecture ribbon.
Select the **Stair by Component** tool from the Circulation panel.

13. ▼ Edit Type | Select **Edit Type** from the Properties pane.

14.
Type Parameters	
Parameter	
Calculation Rules	
Maximum Riser Height	0' 8"
Minimum Tread Depth	0' 11"
Minimum Run Width	4' 0"
Calculation Rules	

Change the Maximum Riser Height to **8"**.
Set the stair width is set to **4' 0"**.
Press **OK**.

15.
Dimensions	
Desired Number of Risers	16
Actual Number of Risers	1
Actual Riser Height	0' 7 1/2"
Actual Tread Depth	0' 11"
Tread/Riser Start Number	1

Set the desired number of risers to **16**.

16. 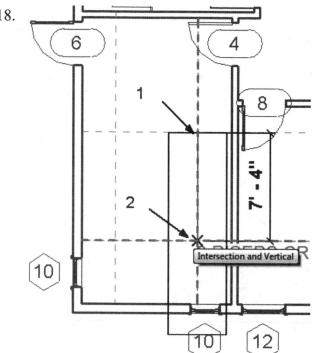 Select the **Run** tool from the Draw panel on the ribbon.

17. Set the Location line to **Run: Center** on the Options bar.

18. Pick the first intersection point indicated for the start of the run. Pick the second point indicated.

19. Moving clockwise, select the lower intersection point #3 and then the upper intersection point #4.

20. 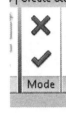 8 U-shaped stairs are placed.

16 RISERS CREATED

21. Select the **Green Check** on the Mode panel to **Finish Stairs**.

Dimensions	
Width	4' 0"
Desired Number of Risers	16
Actual Number of Risers	16
Actual Riser Height	0' 7 1/2"
Actual Tread Depth	0' 11"

22. Select the stairs.

On the Properties pane:
Locate the value for the Actual Tread Depth.
Locate the value for the Actual Riser Height.

23. Close without saving.

Exercise 6-4

Railings

Drawing Name: **m_railings.rvt**
Estimated Time to Completion: 15 Minutes

Scope
Place a railing using Sketch.

Solution

1.

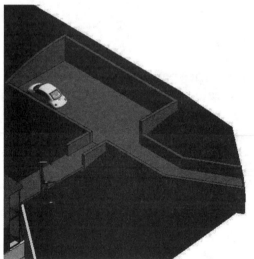

Activate the **Parking Area 3D view**.

2.

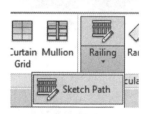

Activate the Architecture ribbon.
Select **Railing→Sketch Path**.

3.

Select **Pick New Host** on the Tools panel.

4.

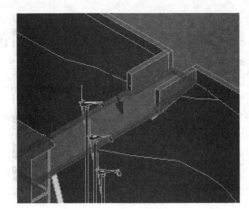

Select the concrete walkway as the host for the railings.

5.

Verify that the Railing type in the Type Selector is set to **SH_1100mm**.

6.

Select the **Pick** tool from the Draw panel.

7.

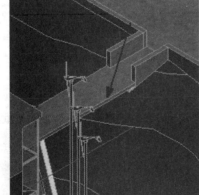

Select the edge of the walkway.

Click the green check to complete the railing.

8.

The railing is placed.

9.

Railings (1)	⌄ 🔡 Edit Type
Constraints	☆
Base Level	
Base Offset	0.0
Offset from Path	0.0
Dimensions	☆
Length	8960.0
Identity Data	☆

Select the railing.
Note that the length of the railing is displayed in the Properties pane.

10. Select **Railing→Sketch Path**.

Curtain Grid   Mullion   Railing   Ra

Sketch Path

11. Select **Pick New Host** on the Tools panel.

Pick New Host   Edit Joins
Tools

12.

Select the concrete walkway as the host for the railings.

13. Railing SH_1100mm

Verify that the Railing type in the Type Selector is set to **SH_1100mm**.

14. Select the **Pick** tool from the Draw panel.

15.

Select the other side edge of the walkway.

Mode

Click the green check to complete the railing.

16. The railing is placed.

17. Save as *ex6-6.rvt*.

Exercise 6-5
Creating a Roof by Footprint

Drawing Name: **i_roofs.rvt**
Estimated Time to Completion: 15 Minutes

Scope
Create a roof.
Determine the roof's volume.

Solution

1. Activate the **T.O. Parapet** floor plan.

   Floor Plans
       Ground Floor
       Lower Roof
       Main Floor
       Main Roof
       Site
       T. O. Footing
       T. O. Parapet
   Ceiling Plans

2. 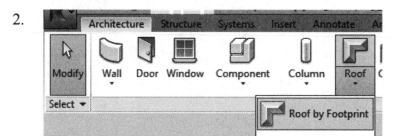 Select **Roof by Footprint** from the ribbon.

3. Select **Steel Truss - Insulation on Metal Deck - EPDM** using the Type Selector on the Properties panel.

4. 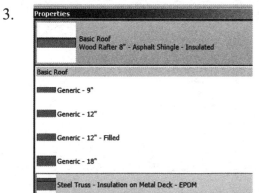 Select **Pick Walls** mode.

5.

On the Options bar:
Uncheck **Defines slope**.
Set the Overhang to **0'-0"**.
Uncheck **Extend to Wall Core**.

6.

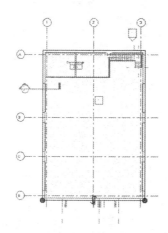

Select all the exterior walls.

7.

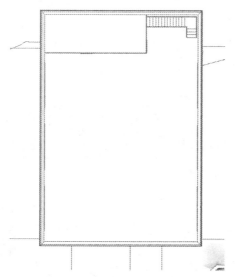

Use the **Trim** tool from the Modify panel to create a closed boundary, if necessary.

8.

Align the boundary lines with the exterior face of each wall.

9. 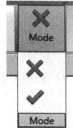 Select the **Green Check** on the Model panel to **Finish Roof**.

10. Select the roof.

In the Properties panel:
Note the volume of the roof.
The volume should be 6631.09. CF.

If your volume is not the same, repeat the exercise.

11. Close without saving.

Exercise 6-6
Modifying a Roof Join

Drawing Name: **roof-join.rvt**
Estimated Time to Completion: 30 Minutes

Thanks to Solomon Smith and Wil Wiens!

Scope
Adding a roof join
Modifying a roof slope
Modifying a roof footprint
Modifying a wall profile

Solution

1.   3D Views   {3D}   Activate the {**3D**} view.

2.   Select the big roof.

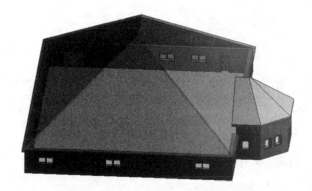

3.

Dimensions	
Slope	9" / 12"
Thickness	0' 8 3/8"
Volume	7559.87 CF
Area	10832.05 SF
Identity Data	

In the Properties panel:
Note the slope is set to **9/12**.

4.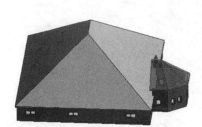
    Select the little roof.

5.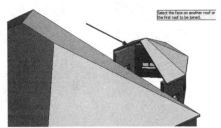
    In the Properties panel:
    Note the slope is set to **9/12**.

6.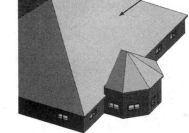
    Activate the **Modify** Ribbon.
    Select **Join/Unjoin Roof** in the Geometry panel.

7. Select the exposed edge of the smaller roof.

    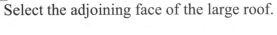

8. Select the adjoining face of the large roof.

9.
    The small roof extends to join the large roof.

10.    Select the small roof.

11.    Change the slope to **6"/12".**

12. 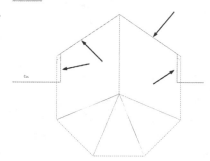   Select the large roof.

13.   Select **Edit Footprint** from the ribbon.

14. 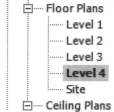   Activate **Level 4.**

15.   Select the **Pick Line** tool.

16.   Select the overlapping edges of the bay roof.
The two vertical lines should be offset in 1' -0".

17.   Use the TRIM and DIVIDE tools to clean up the roof outline.

18.   Select **Green Check** to finish the roof.

19.   Select the **West** elevation.

- Elevations (Building Elevation)
  - East
  - North
  - South
  - **West**

20.   Select the wall indicated.

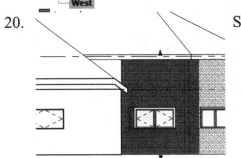

21.   Select **Edit Profile** from the ribbon.

Edit Profile   Reset Profile

Mode

22.   Select the **Pick Line** tool.

Draw

23.   Select the roof edge.

24. 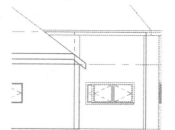 Use the TRIM tool to eliminate the overlapping edges.

25. Select the **Green Check**.

26.  Switch to a 3D view to inspect the west side of the wall.

27. Select the **East** elevation.

28. 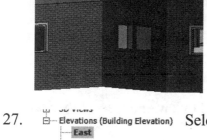 Select the wall indicated.

29.  Select **Edit Profile** from the ribbon.

30.  Select the **Pick Line** tool.

31. 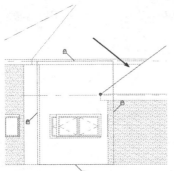 Select the roof edge.

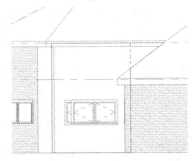

32. Use the TRIM tool to eliminate the overlapping edges.

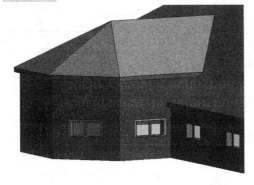

33. Select the **Green Check**.

34. 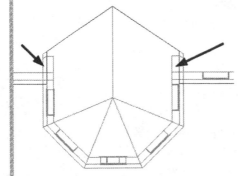 Switch to a 3D view and inspect the walls and roofs.

35. To eliminate any gaps:
Switch to a wire frame display.
Switch to a top view.
Align the boundary of the large roof with the exterior face of the walls indicated.

36. Save as ex6-11.rvt.

Exercise 6-7
Modifying Assembled Stairs

Drawing Name: **modifying_stairs.rvt**
Estimated Time to Completion: 20 Minutes

Scope
Edit an Assembled Stair
Control the number of risers
Modify a landing

Solution

1.   Activate the **{3D} view**.

2.   Select the Stairs.

    Be sure you select the stairs, not the railing to landing. You should see Assembled Stair on the Properties panel if it is selected properly.

    Assembled Stair
    MTL Pan (C 9x15)

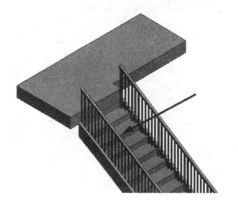

Click on **Edit Stairs** on the ribbon.

The view display changes.
Note that the railings and stringers visibility have been turned off so you can focus on the stairs.

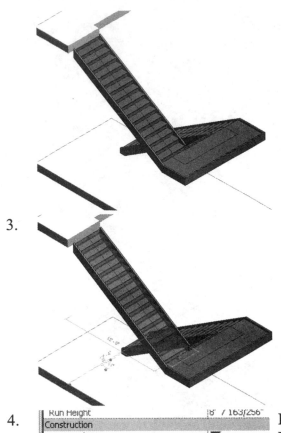

3.　Select the top run.

Note you can control the characteristics of each section of the stair (runs and landings).

4.

In the Properties pane:
Uncheck **End with Riser**.

5.

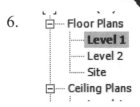

Note that the stair updates.
Left click in the display window to release the stair selection.

Note that we are still in Edit Stair mode - check the ribbon!

6.　Activate **Level 1**.

7.

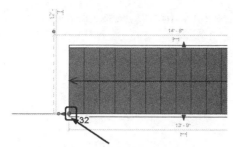

Select the top run again.
Select the round bubble indicated not the arrow.

8.

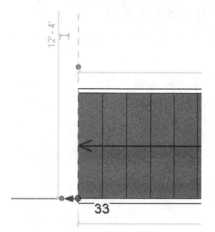

Drag the filled dot handle to align with the floor-edge.

Note that the riser count updates.

9.

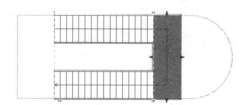

Select the Landing.
This activates shape handles.

10.

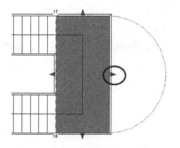

Using the left shape handle extend the landing so it is aligned with the end points of the arc on the floor.

11.

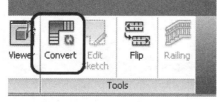

With the landing still selected/highlighted:
Select the **Convert** button on the ribbon.

12.

Stair - Convert to Custom

Converting an automatic landing or a run
component to a custom sketch-based
component is irreversible. You can edit the
sketch of a custom component .

☐ Do not show this message again   Close

You will see a warning message that the
landing will become a sketch-based
component.
Press **Close**.

13.

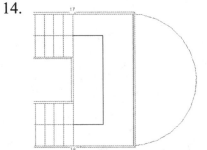

Click on **Edit Sketch** on the ribbon.

14.

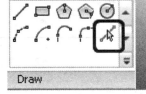

The landing is now in sketch mode.

15.

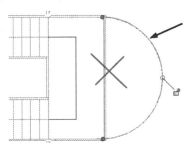

Select the **Pick** tool.

16.

Pick the arc which is part of the floor.
Delete the right vertical line.
Trim the boundary so the horizontal lines are tangent to
the arc.

Your boundary should look like this.

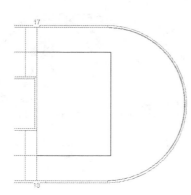

 Select the green check.

If you get an error message, use the TRIM command to clean up the points where the arc connects with the lines.

17. Select green check to finish editing the stairs.

18. 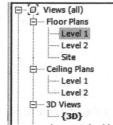 Activate the **{3D} view**.

19. 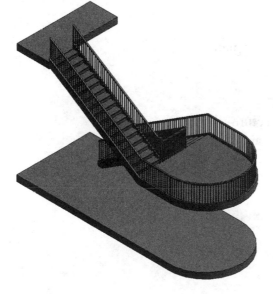 Inspect the modified stairs.

20. Save as *ex6-4.rvt*.

Exercise 6-8
Creating a Stair by Component

Drawing Name: **component_stairs.rvt**
Estimated Time to Completion: 45 Minutes

Thanks to Robert Manna, Krista Manna, and David Light!

Scope
Create a Stair by Component
Modify the Width
Modify a run sketch
Re-position a run
Re-position a landing
Modify a floor's sketch

The walls in this exercise have been set to have transparency. This is done using Visibility/Graphics overrides. An exercise on how to do this is in Lesson 9- Exercise 9-7.

Solution

1. Activate the **Level 1** floor plan.

2. Activate the Architecture ribbon.
    Select the **Stair→Stair by Component** tool.

3. Select the **Railing** button on the ribbon.

4.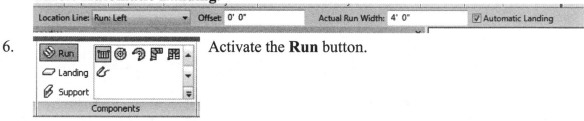

   Select **Guardrail- Pipe (no balusters)** from the drop-down list.
   Press **OK**.

5. On the Options bar: Set the Location Line to **Run:Left**. Set the Offset to **0' 0"**.
   Enable **Automatic Landing**.

Location Line: Run: Left	Offset: 0' 0"	Actual Run Width: 4' 0"	☑ Automatic Landing

6. Activate the **Run** button.

7.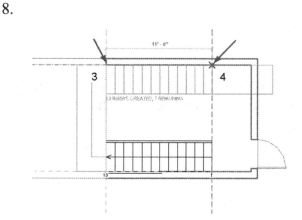

   Start the run at the intersection of the right reference plane and the inside finish face of the bottom horizontal wall (1).

   Then pick at the intersection of the left reference plane and the inside finish face of the bottom horizontal wall (2).

8. Continue the run by picking the intersection of the left reference plane and the inside finish face of the top horizontal wall (3).

   Next pick the intersection of the right reference plane and the inside finish face of the top horizontal wall (4).

9. 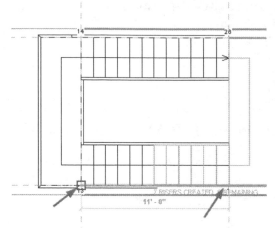 Place a third run on top of the first run using points 1 and 2. (Repeat step 6)

The third run will display as light grey lines.

10. 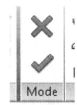 Click on the **Green Check** on the ribbon to complete the stairs.

11. You may see some warnings about errors. Review the errors – we will be fixing them. Close the error dialog box by left clicking on the x in the upper right corner.

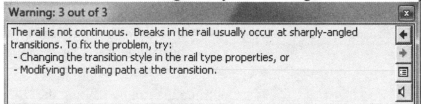

12. 3D Views   Activate a **{3D}** view.
    {3D}

13. 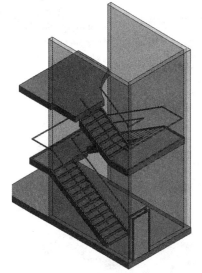 You can see there are some necessary tweaks. Select the stairs so they are highlighted.

14. Select the stairs and select **Edit Stairs** on the ribbon.

Edit
Stairs
Edit

15. Floor Plans | Activate **Level 1**.
Level 1
Level 2
Site

16. Select the second run.

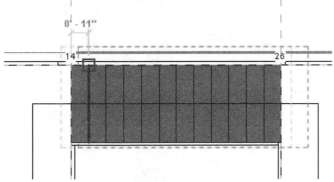

17. Select the **Move** tool.
Enable **Constrain** on the Options bar.
This sets ORTHO on.

Clipboard Geometry

☑ Constrain ☐ Disjoin ☐ Multiple

18. Move the run 11" to the right.

0' - 11"

19. Warning

Landing depth is less than run width.

You will see an error message about the landing.
Close the error dialog.

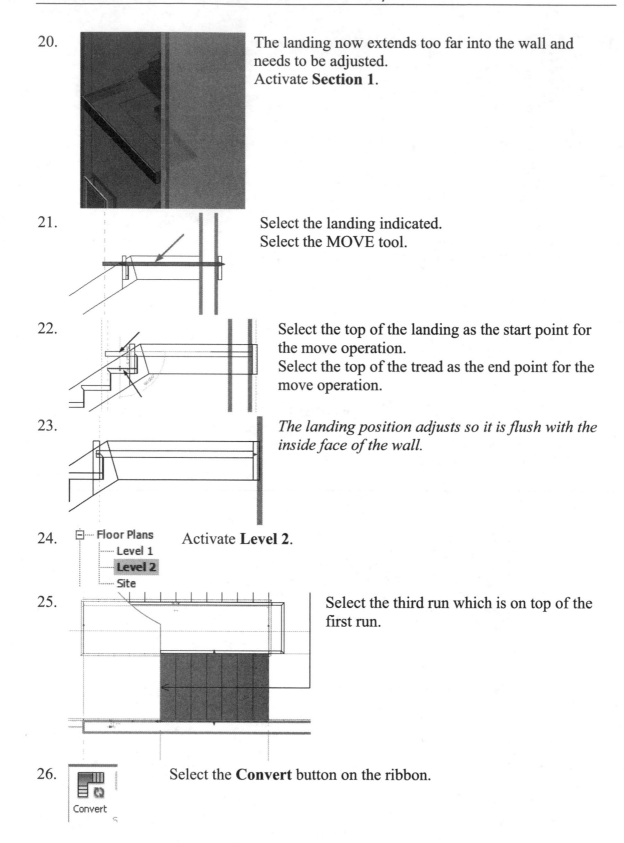

20. The landing now extends too far into the wall and needs to be adjusted.
Activate **Section 1**.

21. Select the landing indicated.
Select the MOVE tool.

22. Select the top of the landing as the start point for the move operation.
Select the top of the tread as the end point for the move operation.

23. *The landing position adjusts so it is flush with the inside face of the wall.*

24. Activate **Level 2**.

25. Select the third run which is on top of the first run.

26. Select the **Convert** button on the ribbon.

Convert

27. 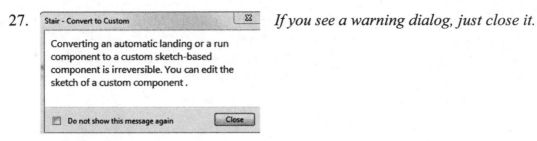 *If you see a warning dialog, just close it.*

Stair - Convert to Custom

Converting an automatic landing or a run component to a custom sketch-based component is irreversible. You can edit the sketch of a custom component.

☐ Do not show this message again   Close

28. Select **Edit Sketch**.

29. Select the **Start-End-Radius arc** tool.

30. 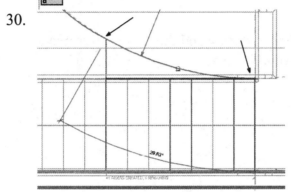 Draw an arc from the endpoint of the landing on the upper left to the finish face of the top horizontal wall endpoint.

31. 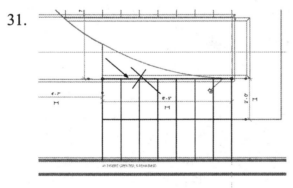 Delete the boundary line indicated by the x-mark.

32. Select the **Extend Multiple** tool.

33. 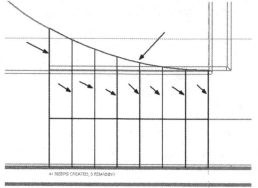 Select the arc as the boundary for the extension.
Then select all the risers to extend them to meet the arc.

34. Select the **Trim** tool.

35. 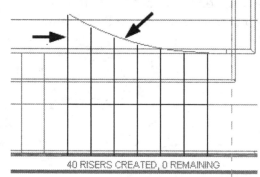 Select the arc and the far left riser line to trim.

36. Select **Green Check** to close the sketch edit.

37. Select **Green Check** to close the stairs edit.

38.
Warning: 1 out of 2	☒
Highlighted lines overlap. Lines may not form closed loops.	← →

Ignore the error warning for the rails. We will fix that later.

39. 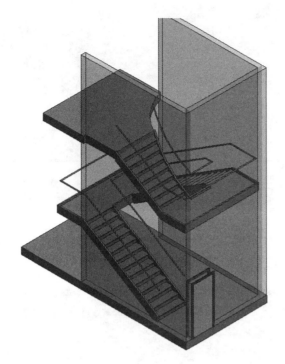 Switch to a 3D view and orbit to inspect your modified stairs.

40. Save as *ex6-5.rvt*.

Exercise 6-9
Baluster Family

Drawing Name: **M_Baluster – Custom 3.rfa**
Estimated Time to Completion: 30 Minutes

Scope
Create a custom baluster family
Create a revolve.
Create an extrusion.
Assign materials.

Solution

1. 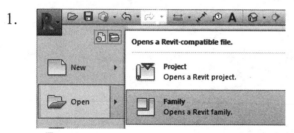   Go to the Applications menu.
   Select **Open→Family**.

2. 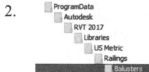   Browse to the *Balusters* folder under *\RVT 2017\Libraries\US Metric\Railings*.

3. File name: | M_Baluster - Custom3   Open *M_Baluster – Custom 3.rfa*.
   Files of type: | Family Files (*.rfa, *.adsk)

4. 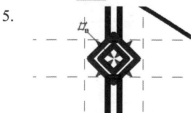   Activate the **Right** elevation.
   - Elevations (Elevation 1)
     - Back
     - Front
     - Left
     - **Right**

5. 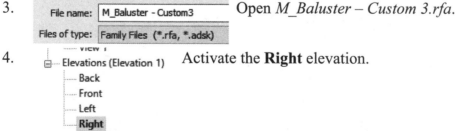   Select the decorative emblem.

6.

Note that the emblem is placed on plane.

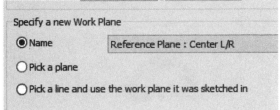

7.

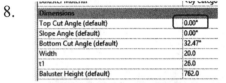

Select **Family Types**.

8.

Dimensions	
Top Cut Angle (default)	0.00°
Slope Angle (default)	0.00°
Bottom Cut Angle (default)	32.47°
Width	20.0
t1	26.0
Baluster Height (default)	762.0

Set the Top Cut Angle to **0.00**.
Press **OK**.

9.

On the Create ribbon: Select **Revolve**.

10.

When the Specify a New Work Plane dialog appears: Enable **Name**. Select **Center L/R**. Press **OK**.

11.

Draw a quarter arc – a vertical line collinear to the center plane, a horizontal line with one end at the center line and one end at the right reference plane, close with a small arc.

12.

Select **Axis Line** and then the **Pick Line** tool

13. 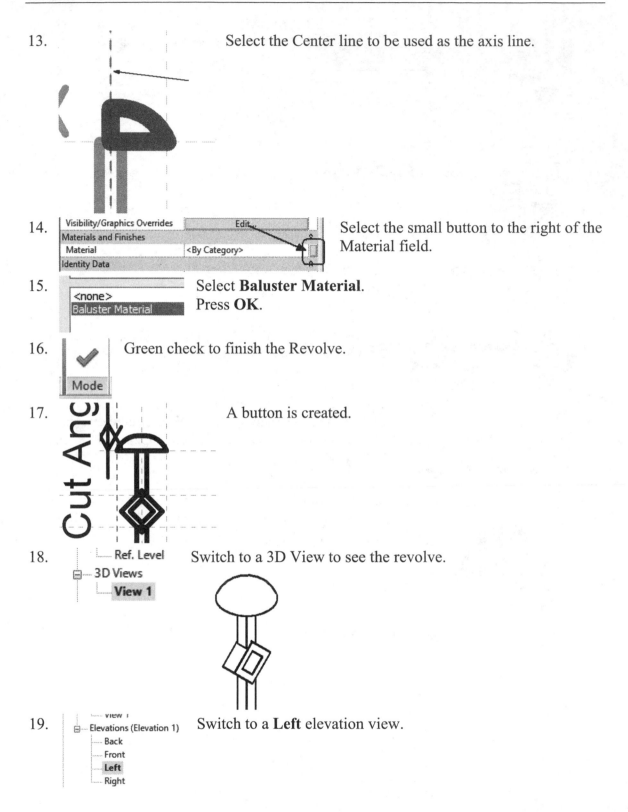 Select the Center line to be used as the axis line.

14.

Visibility/Graphics Overrides	Edit...
Materials and Finishes	
Material	<By Category>
Identity Data	

Select the small button to the right of the Material field.

15.

<none>
Baluster Material

Select **Baluster Material**.
Press **OK**.

16. Green check to finish the Revolve.

Mode

17. A button is created.

Cut Ang

18. Ref. Level
3D Views
View 1

Switch to a 3D View to see the revolve.

19. View 1
Elevations (Elevation 1)
Back
Front
Left
Right

Switch to a **Left** elevation view.

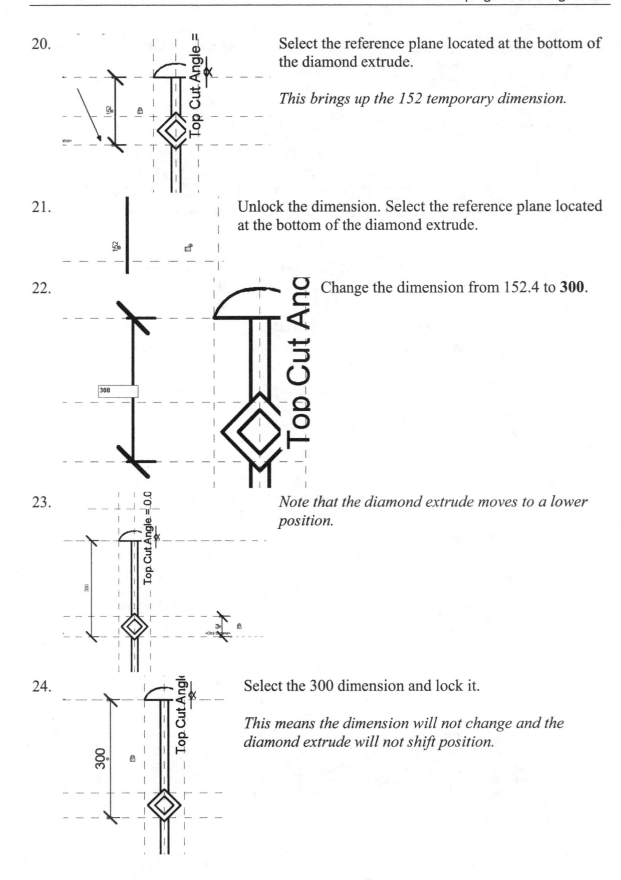

20. Select the reference plane located at the bottom of the diamond extrude.

This brings up the 152 temporary dimension.

21. Unlock the dimension. Select the reference plane located at the bottom of the diamond extrude.

22. Change the dimension from 152.4 to **300**.

23. *Note that the diamond extrude moves to a lower position.*

24. Select the 300 dimension and lock it.

This means the dimension will not change and the diamond extrude will not shift position.

25. 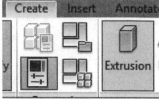 On the Create ribbon: Select **Extrusion**.

26. 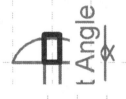 Set the Extrusion End to **13.0**.
Set the Extrusion Start to **-13.0**.
Note the work plane is set to **Center L/R**.
If your properties don't show the correct work plane, use Set Work Plane to change to the correct plane.

27. Select the small button to the right of the Material field.

28. Select **Baluster Material**.
Press **OK**.

29. 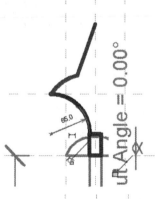 Draw a small rectangle which is aligned to the top and bottom of the button and left and right sides of the baluster.

30. Draw an arc with a radius of 65 with one end located at the left top of the rectangle and one end point coincident to the reference plane.

31. 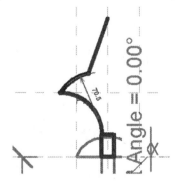 Draw a second arc with a radius of 70.5.

32.

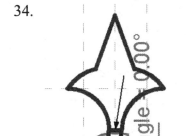

Draw an angled line with an angle of 70°.

33.

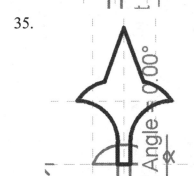

 Use the **Mirror – Pick Axis** tool to mirror the two arcs and line to the other side.

34. Delete the top short line of the rectangle.

35. You have drawn a *fleur de lis* to top your baluster. Green check to complete the extrusion.

36. Switch to a 3D view to inspect your baluster.

37. Save as *M_Baluster – Crescent City.rfa*.

Exercise 6-10
Railing Family

Drawing Name: **m_railing_family.rvt**
Estimated Time to Completion: 30 Minutes

Scope
Create a custom railing system family.
Create global parameters to manage materials.
Place railings on walls.

Solution

1.   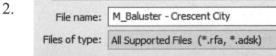   Go to the Insert ribbon.
Select **Load Family**.

2.   Select the *M_Baluster – Crescent City* family.
Press **Open**.

File name:	M_Baluster - Crescent City
Files of type:	All Supported Files (*.rfa, *.adsk)

3.   Select one of the railings placed on the concrete walkway.

4.   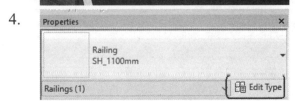   Select **Edit Type** on the Properties panel.

Properties   ×

Railing
SH_1100mm

Railings (1)   Edit Type

5. 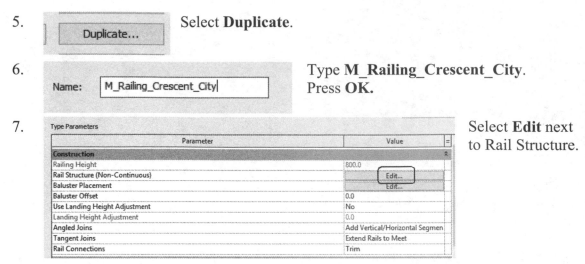 Select **Duplicate**.

6. Type **M_Railing_Crescent_City**.
    Press **OK**.

    Name: | M_Railing_Crescent_City |

7. Select **Edit** next to Rail Structure.

Parameter	Value	=
Construction		⌃
Railing Height	800.0	
Rail Structure (Non-Continuous)	Edit...	
Baluster Placement	Edit...	
Baluster Offset	0.0	
Use Landing Height Adjustment	No	
Landing Height Adjustment	0.0	
Angled Joins	Add Vertical/Horizontal Segmen	
Tangent Joins	Extend Rails to Meet	
Rail Connections	Trim	

8. Delete all the rails except for **New Rail (1)**. Set the Height to **250.0**. Set the Profile to **M_Rectangular Handrail: 50 x 50 mm.** Set the Material to **Iron, Wrought.**
    You will have to import the material into the document.
    Press **OK**.

Family: Railing
Type: M_Railing_Crescent_City

Rails

	Name	Height	Offset	Profile	Material
1	New Rail(1)	250.0	0.0	M_Rectangular Handrail : 50 x 50mm	Iron, Wrought

9. Select **Edit** next to Baluster Placement.

Construction	
Railing Height	800.0
Rail Structure (Non-Continuous)	Edit...
Baluster Placement	Edit...
Baluster Offset	0.0
Use Landing Height Adjustment	No
Landing Height Adjustment	0.0
Angled Joins	Add Vertical/Horizontal Segments
Tangent Joins	Extend Rails to Meet
Rail Connections	Trim
Top Rail	

10. Set the Baluster Family to **M_Baluster – Crescent City : 25 mm**. Set the Host to **Top Rail Element**. Set the Top Offset to **50.0**. Set Distance from Previous to **200.00.**

Main pattern

	Name	Baluster Family	Base	Base offset	Top	Top offset	Dist. from previous	Offset
1	Pattern start	N/A	N/A	N/A	N/A	N/A	N/A	N/A
2	Regular balu	M_Baluster - Crescent Cit	Host	0.0	Top Rail Ele	50.0	200.0	0.0
3	Pattern end	N/A	N/A	N/A	N/A	N/A	0.0	N/A

11. Set the Start Post to **None**. Set the Corner Post to **None**. Set the End Post to **None**. Press **OK**.

Posts

	Name	Baluster Family	Base	Base offset	Top	Top offset	Space	Offset
1	Start Post	None	Host	-247.0	Top Rail Ele	0.0	0.0	0.0
2	Corner Post	None	Host	-247.0	Top Rail Ele	0.0	0.0	0.0
3	End Post	None	Host	-247.0	Top Rail Ele	0.0	0.0	0.0

12.

Top Rail	
Height	800.0
Type	Rectangular - 50x50mm

Set the Top Rail Height to **800.00**. Press **OK**.

13.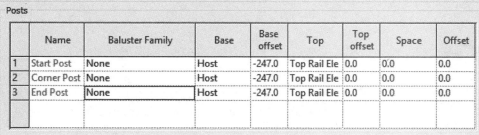

Parking Area
Section Perspective
Solar Analysis
{3D}

Activate a 3D view and orbit to inspect the railing.

14. Select the **Manage** ribbon. Select **Global Parameters**.

15. Select **New Parameter**.

16. **Global Parameter Properties**

Name:
Baluster Material

☐ Reporting Parameter
(Can be used to extract value from a geometric condition and report it in a formula)

Discipline:
Common

Type of parameter:
Material

Group parameter under:
Materials and Finishes

Tooltip description:
<No tooltip description. Edit this parameter to write a custom toolti...

Edit Tooltip...

How do I create global parameters?

OK   Cancel

Type **Baluster Material**. Set the Type of parameter to **Material**. Press **OK**.

17.

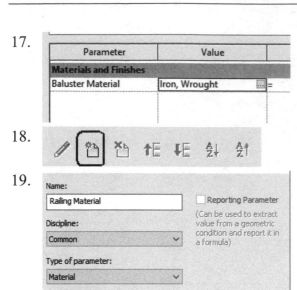

Set the Baluster Material to **Iron, Wrought**.

18.

Select **New Parameter**.

19.

Type **Railing Material**. Set the Type of parameter to **Material**. Press **OK**.

20.

Set the Railing Material to **Iron, Wrought**. Press **OK**.

21.

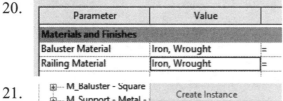

In the Browser:
Locate the Top Rail Type: Rectangular – 50 x 50 mm.
Right click and select **Type Properties**.

22. Assign the Material to **Iron, Wrought**. Press **OK**.

23.  The railing now appears to be a uniform material.

24. 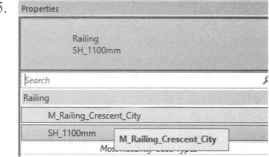 Select the second railing that was placed.

25.  Use the Type Selector to change the railing to **M_Railing_Crescent_City.**

26. Kitchen
    Living Room
    Parking Area
    Section Perspective
    Solar Analysis
    {3D}

    Activate the Parking Area 3D view.

27.   Use the PAN tool to move the view over to the area where the car is parked.

28. 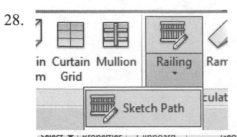   On the Architecture ribbon:
    Select **Railing** →**Sketch Path**.

29. 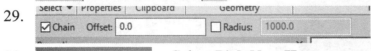   Enable **Chain** on the Options
    bar.

30.    Select **Pick New Host**.

31.    Select the curved wall.

32.    Select the **Pick** tool.

33. 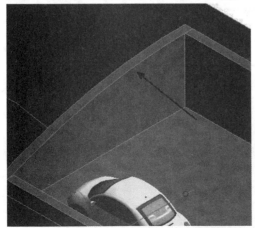   Select the top edge of the curved wall.

34.

Select the two walls indicated.

35.

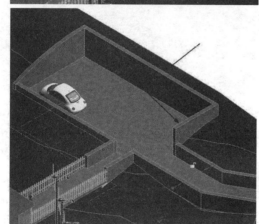

Select the remaining connecting walls.
Do not select the lower walls.

Use the TRIM tool to create a continuous polyline.

36.

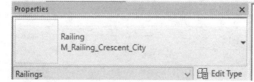

Set the Railing Type to
M_Railing_Crescent_City.

37.

Click the green check to finish.

38.

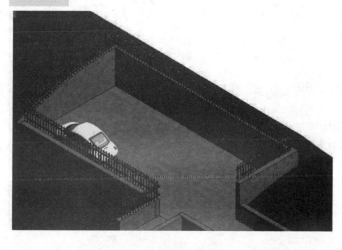

The railing is placed.
Save as *ex6-8.rvt.*

Exercise 6-11
Creating a Roof by Extrusion

Drawing Name: **i_roofs_extrusion.rvt**
Estimated Time to Completion: 30 Minutes

Scope
Create a roof by extrusion.
Modify a roof.

Solution

1. ⊟─ **3D Views**   Activate the **3D view**.
   └─ **{3D}**

2. 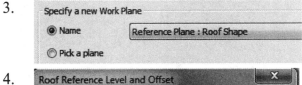   Activate the Architecture ribbon.
   Select **Roof by Extrusion** under the Build panel.

3.    Enable **Name**. Select **Roof Shape** from the list of reference planes.
   Press **OK**.

4. Select **Upper Roof** from the list.
   Press **OK**.

5.    Select the **Show** tool from the Work Plane panel.
   This will display the active work plane.

6.

Right click on the ViewCube's ring.
Select **Orient to a Plane**.

7.

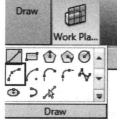

Enable **Name**. Select **Roof Shape** from the
list of reference planes. Press **OK**.

8.

Select the **Start-End-Radius Arc** tool from the Draw panel.

9.

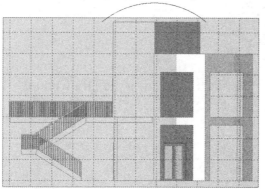

Draw an arc over the building as shown.

10.

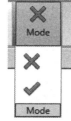

Select the **Green Check** under the Mode panel to finish the roof.

11. Switch to an isometric 3D view.

12. Select the wall below the roof.

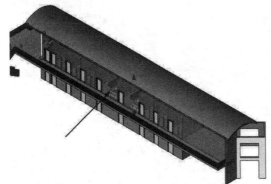

13. 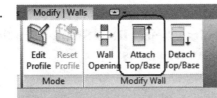 Select **Attach Top/Base** from the Modify Wall panel. Then select the roof. The wall will adjust to meet the roof. Go around the building and attach the appropriate remaining walls.

14. Activate the **Level 3** floor plan.

Floor Plans
Level 1
Level 2
Level 3
Site
T.O.F.

15.  Activate the Architecture ribbon. Select the **Roof By Footprint** tool from the Build panel.

16. 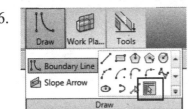 Select the **Pick Walls** tool from the Draw panel.

17. On the Options bar:

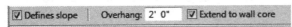

Enable **Defines slope**.
Set the Overhang to **2′ 0″**.
Enable **Extend to wall core**.

18.

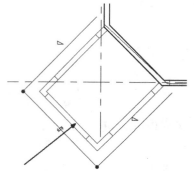

Select the outside edge of the two walls indicated.

19.

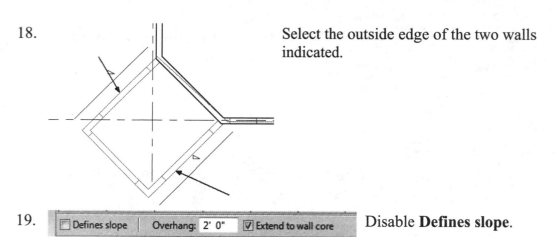

Disable **Defines slope**.

20.

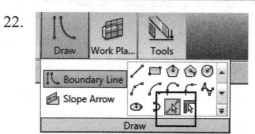

Select the outside edge of the walls indicated.

21.

Set the Overhang to **0′ 0″**.

22.

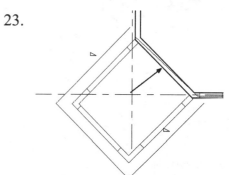

Select the **Pick Line** tool from the Draw panel.

23.

Pick the exterior side of the wall indicated.

24. 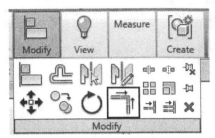 Select the **Trim** tool from the Modify panel.

25. 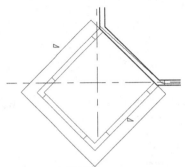 Trim the roof boundaries so they form a closed polygon.

26. 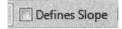 Select **Edit Type** in the Properties pane.

27. Type: Generic - 9"   Set the Type to **Generic- 9″**. Press **OK**.

28. Dimensions / Slope   6" / 12"   Set the Slope to **6″/12″**.

29. 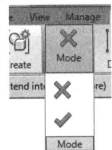 Select the back and front boundary and lines and uncheck **Defines Slope** on the Options bar.

    ☐ Defines Slope

30. Once the sketch is closed and trimmed properly, select the **Green Check** on the Mode panel.

31. 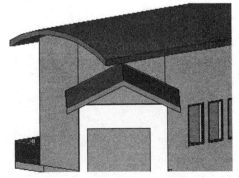 Return to a 3D view.

32.  Select the front wall.

33.  Select Attach Top/Base from the Modify Wall panel.

34. 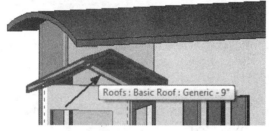 Select the roof to attach the wall.

35. Close without saving.

Exercise 6-12
Creating a Sloped Ceiling

Drawing Name: **i_ceilings.rvt**
Estimated Time to Completion: 30 Minutes

Scope
Create a sloped ceiling.

Solution

1. Ceiling Plans
     ┈ Ground Floor
     ┈ **Ground Floor Admin Wing**
     ┈ Lower Roof
     ┈ Main Floor
     ┈ Main Roof
     ┈ T. O. Footing
     ┈ T. O. Parapet

    Activate the **Ground Floor Admin Wing** ceiling plan.

    A sloped ceiling will be placed in the room indicated.

2. Select the **Ceiling** tool from the Build panel on the Architecture ribbon.

3. Select **Sketch Ceiling** from the Ceiling panel.

4. 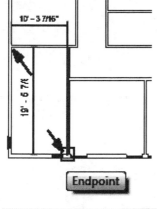   Select the **Rectangle** tool from the Draw panel.

5. Draw a rectangle inside the room by selecting opposing corners.

6. Select the **Slope Arrow** tool from the Draw panel to place a slope.

7. Using the Draw tool:

   Select the Midpoint at the top of the rectangle.

8. Bring the arrow down to indicate the direction of the slope. Pick to place the endpoint so it is coincident with the lower wall.

9. 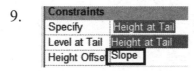   On the Properties pane:
   Under Specify: Select **Slope**.

10.

Parameter	
Constraints	
Specify	Slope
Level at Tail	Default
Height Offset	0' 6"
Level at Hea	Default
Height Offset	0' 0"
Dimensions	
Slope	2" / 12"
Length	19' 5 7/8"

Set the Height Offset at Tail to **6″**.
Set the Slope to **2″/12″**.
Left click in the drawing window to release the selection.

11.

Properties

Compound Ceiling
2' x 2' ACT System

Basic Ceiling

   Generic

Compound Ceiling

   2' x 2' ACT System

   2' x 4' ACT System

   GWB on Mtl. Stud

In the Properties pane:
Use the Type Selector to set the ceiling type to **GWB on Mtl. Stud**.

12.

Mode

Mode

Select the **Green Check** on the Mode panel to **Finish Ceiling**.

13.

The ceiling does not display a hatch pattern by default.

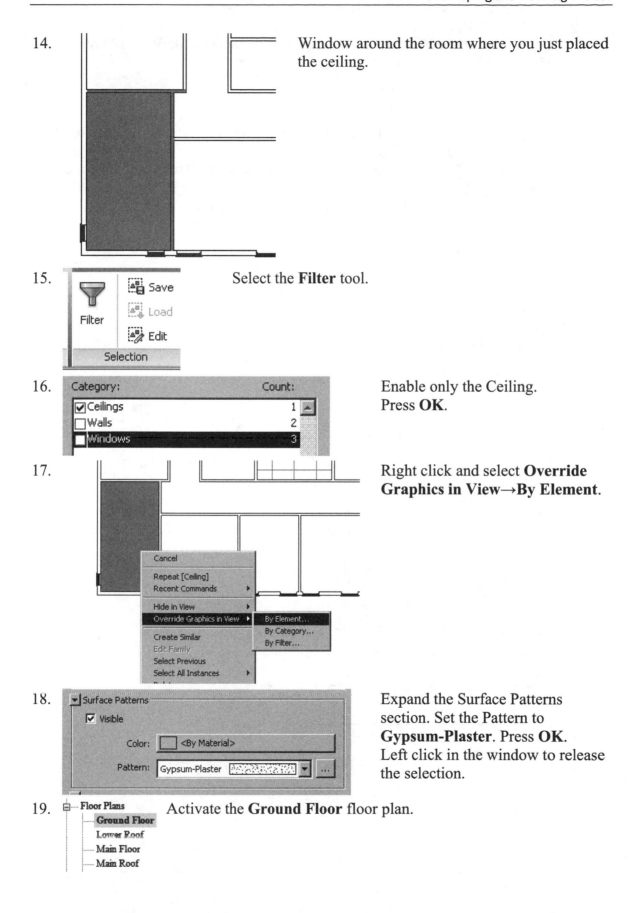

14.   Window around the room where you just placed the ceiling.

15.   Select the **Filter** tool.

16.   Enable only the Ceiling.
Press **OK**.

17.   Right click and select **Override Graphics in View→By Element**.

18.   Expand the Surface Patterns section. Set the Pattern to **Gypsum-Plaster**. Press **OK**. Left click in the window to release the selection.

19.   Activate the **Ground Floor** floor plan.

20. Select the **Section** tool from the Create panel on the View ribbon.

Section

21. 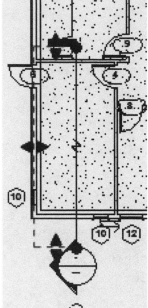 Place the section so it intersects the room with the sloped ceiling.
Left click below the room at door 10 to start the section line.
Left click above door 6 to end the section line.
Double click on the section head bubble to activate the section view.

22. 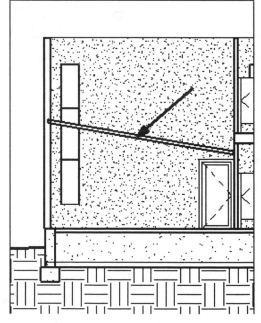 The sloped ceiling is shown.
Select the ceiling.

23.

Constraints	
Level	Ground Floor
Height Offset From Level	7' 0"
Room Bounding	☑
Dimensions	
Slope	1/4" / 12"
Perimeter	59' 6 5/8"
Area	200.52 SF
Volume	71.02 CF
Identity Data	
Comments	
Mark	
Phasing	
Phase Created	New Construction
Phase Demolished	None

In the Properties pane:
Set the Height Offset to **7' 0"**.
Set the Slope to **1/4"/12"**.
Press **Apply**.

24.

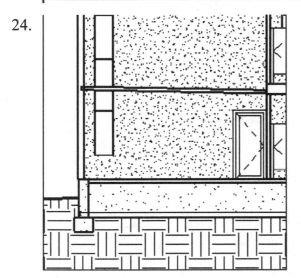

The ceiling updates with the new values.

25. Close without saving.

Certified User Practice Exam

1. To create a partial ceiling:

   A. Use a sketch
   B. Create an area first
   C. Use the partial ceiling tool
   D. Create a modified ceiling

2. The boundary of a floor is defined by:

   A. Model Lines
   B. Reference Lines
   C. Slab Edges
   D. Sketch Lines
   E. Drafting Lines

3. For the stairs shown, the vertical green lines represent:

   A. Risers
   B. Boundary Lines
   C. Stringers
   D. Run
   E. Railings

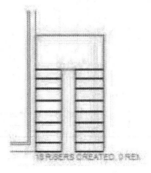

4. **True or False**: Stairs can be placed on multiple levels at a time.

5. **True or False**: You can change the direction of stairs using the flip arrows.

6. Identify the type of roof:

   A. Gable
   B. Hip
   C. Tar and Gravel
   D. Flat
   E. Shingled

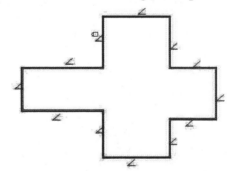

7. Name the three methods used to place a roof:

   A. By face
   B. By profile
   C. By footprint
   D. By extrusion
   E. By sketch

8. **True or False**: The extrusion of a roof can extend only in a positive direction from the selected work plane.

9. **True or False**: When attaching a wall to a roof, the profile of the wall does not change.

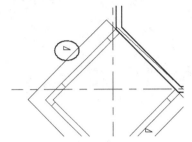

10. The symbol circled indicates:

    A. Slope
    B. Top Constraint
    C. Roof Boundary
    D. Gutter

11. To place a railing:

    A. Draw a path
    B. Select a host element
    C. Click to place
    D. Work in section view

12. You can change a Generic 12" Floor to a Wood Joist 10" Floor using the:

    A. Properties filter
    B. Options Bar
    C. Type Selector
    D. Modify ribbon

13. To create a stair landing, use the _____ tool.

    A. Run
    B. Boundary
    C. Riser
    D. Landing

14. To attach the top of a wall to a roof:

    A. Drag the top of the wall by grips
    B. Select the wall, check on Attach Top/Base and select roof
    C. Define wall attachments in instance properties
    D. Use the ALIGN tool

15. Railings can be hosted by the following: (select all that apply)
    A. Slab edges
    B. Roofs
    C. Walls
    D. Stairs
    E. Ramps
    F. Floors

Answers:
  1) A; 2) D; 3) B; 4) True; 5) True; 6) B; 7) A, C & D; 8) False; 9) False; 10) A) Slope; 11) B; 12) C; 13) B; 14) B; 15) A, B, C,
D, E, and F

Certified Professional Practice Exam

1. Component-based stairs use: (Select the TWO best answers)

    A. Common components
    B. Custom Sketched Components
    C. Dimensions
    D. Type Properties

2. A component-based stair can consist of all of the following EXCEPT:

    A. Runs
    B. Landings
    C. Supports
    D. Railings
    E. Tags

3. A Stair Run can be all the following shapes EXCEPT:

    A. Straight
    B. Spiral
    C. U-Shaped
    D. Closed

4. **True or False**: A stair created by components cannot be added to an assembly.

5. The shapes of rails and balusters are determined by _____, which must be loaded in the project.

    A. Profile families
    B. Type Properties
    C. System families
    D. In-place masses

Answers:
1) A & B; 2) E; 3) D; 4) T; 5) A

Detailing and Drafting

This lesson addresses the following certification exam questions:

- Callouts
- Tags
- Detail Views
- Text
- Callout view of a Section
- Drafting View
- Revision Clouds
- Revision Schedules
- Grid Guides

A callout is a view that is placed in a plan, section, detail or elevation view to create a more detailed view of part of the building model. The area that the callout defines in the view is called the callout bubble. The callout bubble has a leader or connecting line to a callout head which displays the detail number and sheet number. All these parts – callout bubble, callout head, and callout leader – are referred to as a callout tag.

When you create a callout, a new view is automatically created. There are two types of callout views: Callout in Parent View and Callout in Detail View.

A detail view is a view of a specific area of a plan, elevation or section view. This view provides a greater level of detail at a larger scale than the parent view. Detail views can be linked to callouts.

Drafting views are 2D views that depict a small area of the building model. These views do not contain any model elements at all. Users can create a library of drafting views to depict foundation work, stairs and railings, wall framing, etc., and use them across multiple projects.

Revision clouds and revision schedules are used to track changes to a project. Users can enter the changes into a project and then issue the changes to a sheet once the changes are made.

Exercise 7-1

Creating Drafting Views

Drawing Name: **c_libraries_and_details.rvt**
Estimated Time to Completion: 30 Minutes

Scope
Use an existing AutoCAD detail drawing and convert it into a Revit drafting view.
Add a callout to a section view.
Add a section view with callout to a sheet.
Add a drafting view to a sheet.

Solution

1. Activate the **View** ribbon.
    Select **Drafting View** from the Create panel.

    Drafting
    View

2.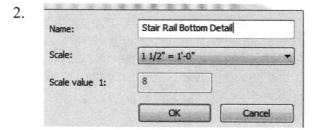
    Name the new Drafting View:
    Stair Rail Bottom Detail.
    Set the scale to 1 ½″ = 1′-0″
    Press **OK**.

Name:	Stair Rail Bottom Detail
Scale:	1 1/2" = 1'-0"
Scale value 1:	8

    OK   Cancel

3. ⋯⋯ vvest
    ⊟⋯ Sections (Building Section)
    ⋮ ⋯ Stair Section
    ⊟⋯ Drafting Views (Detail)
    ⋮ ⋯ **Stair Rail Bottom Detail**

    In the Project Browser, note that the Drafting Views
    category has been added and the new view is listed.
    The window will open to the new view which is empty.

4. Activate the **Insert** ribbon.
    Select **Import CAD** from the Import panel.

    CAD
    Import
    CAD

5. Locate the *c_detail_bottom.dwg* file. Set Colors to **Invert**. Set positioning to **Auto-Center to Center**. Press **Open**.

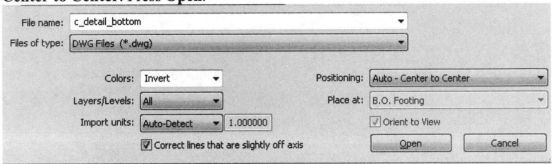

6.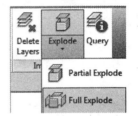

   Select the CAD data. It is a single image.
   Right click and select **Full Explode**.
   This will convert the data to Revit entities.

   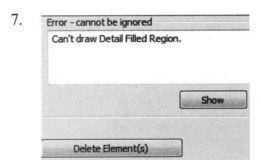

   You can also do a Full Explode from the ribbon.

7.

   Error - cannot be ignored

   Can't draw Detail Filled Region.

   Show

   Delete Element(s)

   An error message may appear.
   Select **Delete Element(s)**.

   Revit was unable to convert the detail filled region, but you can add that later.

8. West
   Sections (Building
   Stair Section

   Next we add a callout for the drafting view.
   Activate the **Stair Section** under Sections.

9.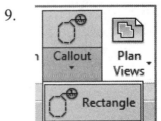

   Activate the **View** ribbon.
   Select the **Callout Rectangle** tool from the Create panel.

10.

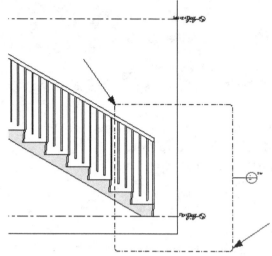

Place a check on the **Reference other view** box. In the drop-down list, select the **Stair Rail Bottom Detail** for the view to be referenced.

11.

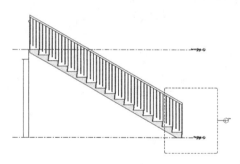

Place the callout outline by picking in the two locations indicated.

12.

Some users don't like to see the viewport outline. This is the rectangle around the view.

Use the **Hide Crop Region** toggle on the View Control bar to hide the viewport outline.

13.

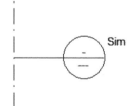

The rectangular viewport is no longer visible.

14.

Zoom into the callout bubble.

The bubble data is blank until the drafting view is added to a sheet.

15. Sheets (all)
    A1 - First Floor Plan
    A2 - Second Floor Plan
    A3 - First Floor Reflected Ceiling Plan
    A4 - Second Floor Reflected Ceiling Plan
    A5 - Roof Plan
    A6 - Stair Section
    A7 - Stair Details

Activate the **Stair Section** sheet.

16.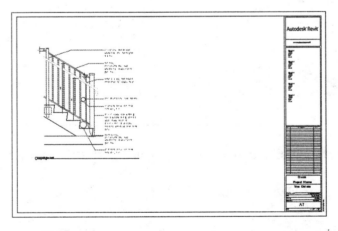

Drag and drop the Stair Section from the browser onto the sheet.

17. Sheets (all)
    A1 - First Floor Plan
    A2 - Second Floor Plan
    A3 - First Floor Reflected Ceiling Plan
    A4 - Second Floor Reflected Ceiling Plan
    A5 - Roof Plan
    A6 - Stair Section
    A7 - Stair Details

Activate the **Stair Details** sheet.

18.

Drag and drop the Drafting View for the bottom of the stairs onto the sheet.

19. West
    Sections (Building Sec
    Stair Section

Activate the **Stair Section** view under Sections.

20.
Zoom into the callout bubble.
Note that the callout is now filled in with the sheet information.
Double left click on the bubble.

21. Drafting Views (Detail)
    Stair Rail Bottom Detail

    This navigates the user to the drafting view.

22. Sections (Building Section)
    Stair Section

    Activate the **Stair Section** view under Sections.

23. Select the callout handle.

    Sim
    1
    A7

24. Edit Type   Select **Edit Type** in the Properties pane.

25.

Parameter	Value
Graphics	
Callout Tag	Callout Head w 1/8" Corner Radius
Section Tag	Section Head - No Arrow, Section Tail - Filled Horizontal
Reference Label	
Identity Data	

Delete the word **Sim** in the Reference Label field.
Press **OK**.

26.

Identity Data	
View Template	<None>
View Name	Stair Rail Bottom Detail
Dependency	Independent
Title on Sheet	
Sheet Number	A7
Sheet Name	Stair Details
Referencing Sheet	A6
Referencing Detail	1

Activate the **Stair Rail Bottom Detail** view.

In the Properties pane:
Note that the Referencing Sheet is listed under the Identity Data

None of the data is editable in this dialog.

27. Save as *ex7-1.rvt*.

Extra Challenge

Create a callout for the top of the stairs using c_detail_top.dwg as the drafting view.

Add the new drafting view to the Stair Details sheet.

1. Create a new drafting view.
2. Import the CAD file.
3. Add the callout to the top of the stairs on the section view.
4. Drag the new drafting view to the Stair Details sheet.

Exercise 7-2
Reassociate a Callout

Drawing Name: **callout_redefine.rvt**
Estimated Time to Completion: 10 Minutes

Scope
Change the drafting view for a callout

Solution

1.
```
    {3D}
Elevations (Building Elevation)
    East
    North
    South
    West
```
Activate the **North Elevation** view.

2.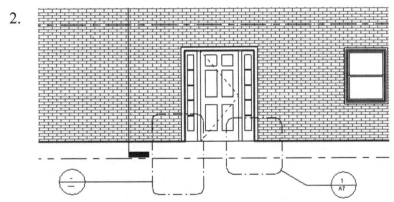
There are two call-outs on the exterior door.
The callout on the right references a sheet with a drafting view.
The callout on the left has no sheet or view references.

3. Select the left callout with the blank bubble. Note that the Reference other view is grayed out. If you do not assign a reference view when you place a callout, you can't create the link later.

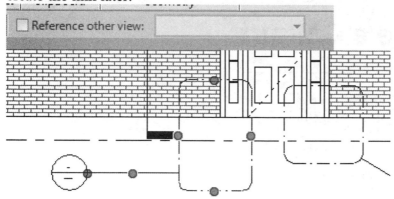

4. 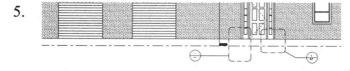 Open Sheet **A7 - Stair Details**.
This is the sheet referenced by the callout.

5. 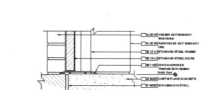 Open Sheet **A8- Door Details**.
The callout should be pointing to the Exterior Door Threshold Detail.

6.

Cancel	
Repeat [Delete]	
Recent Commands	▶
Activate View	
Hide in View	▶

Right click on the elevation view on the sheet.
Select **Activate View**.

7. 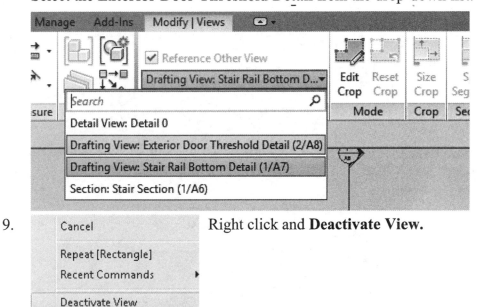 Select the right callout.

8. On the ribbon:
   Select the **Exterior Door Threshold Detail** from the drop-down list.

9. Right click and **Deactivate View.**

10. Note how the callout updates with the correct sheet and view number.

11. Close without saving.

Exercise 7-3

Filled Regions

Drawing Name: **ex7-1.rvt**
Estimated Time to Completion: 25 Minutes

Scope
Use Filled Regions to modify elements in a drafting view

Solution

1.
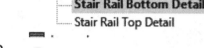

Activate the **Stair Rail Bottom Detail** view.

2.
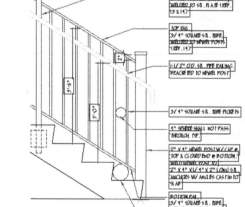

Select all the text.

Use the Type Selector in the Properties panel to change the text to **1/4" City Blueprint**.

3.
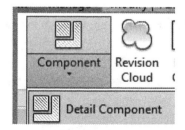

Activate the Annotate ribbon.
Select **Detail Component**.

4. 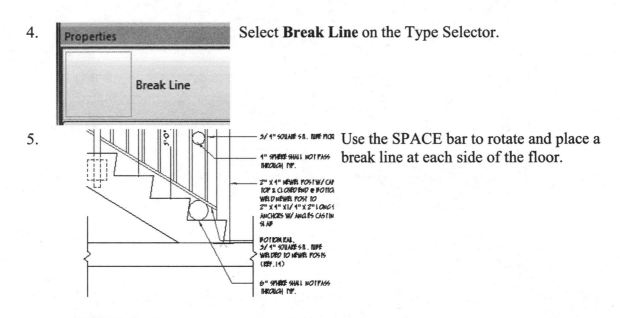 Select **Break Line** on the Type Selector.

5. Use the SPACE bar to rotate and place a break line at each side of the floor.

6. Select the **Filled Region** tool.

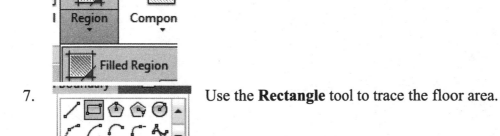

7. Use the **Rectangle** tool to trace the floor area.

8. Select the **Floor Filled Region** from the Type Selector.

9. **Green Check** to finish.

10.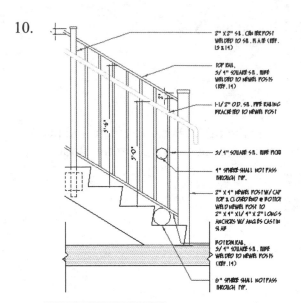

The filled region is displayed.

11. Select the **Filled Region** tool.

12. Use the **Pick Line** tool to trace the two circles.

13.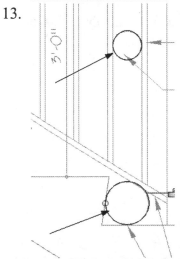

Select the two circles.

14. Select the **DO NOT PASS SPHERE** filled region from the Type Selector.

Filled region
DO NOT PASS SPHERE

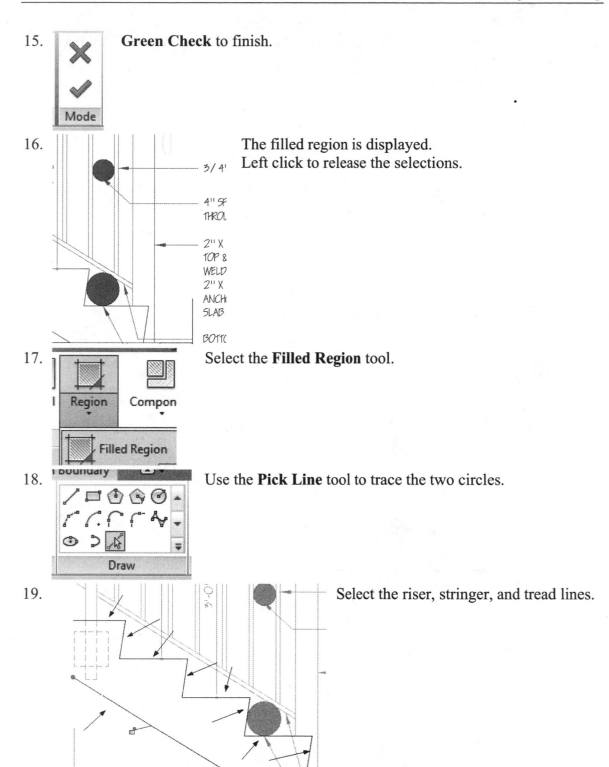

15. **Green Check** to finish.

16. The filled region is displayed.
     Left click to release the selections.

     3/4'

     4" SE
     THROL

     2" X
     TOP 8
     WELD
     2" X
     ANCH
     SLAB

     BOTTC

17. Select the **Filled Region** tool.

18. Use the **Pick Line** tool to trace the two circles.

19. Select the riser, stringer, and tread lines.

20. Use the LINE tool to close the filled region.

21. Select the **Treads** Filled Region from the Type Selector.

Filled region
Treads

22.  **Green Check** to finish.

23. The filled region is displayed.

3/ 4" SQUARE

4" SPHERE SHA
THROUGH. TYP.

2" X 4" NEWE
TOP & CLOSED
WELD NEWEL F
2" X 4" X 1/ ·
ANCHORS W/ .
SLAB

BOTTOM RAIL.
3/ 4" SQUARE
WELDED TO NE
(REF. 14)

6" SPHERE SHA
THROUGH. TYP.

24. Save as *ex7-3.rvt*.

Exercise 7-4

Adding Tags

Drawing Name: **i_dimensions.rvt**
Estimated Time to Completion: 15 Minutes

Scope
Add room tags to a view.
Duplicate view with Detailing
Duplicate view as dependent
Change the room tags assigned type.

Solution

1.  Activate the **Ground Floor Admin Wing**.

2. 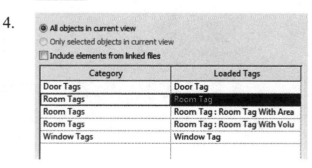 Mouse over the different rooms and note that room elements have been placed in each room.

3. Activate the **Annotate** ribbon.
Select the **Tag All** tool from the Tag panel.

4. 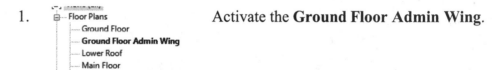 Select the **Room Tag**.
Press **OK**.

5. Press **Yes** if you see this dialog.

> **Tag Visibility Enabled** ✕
>
> Visibility of Room Tags is turned off in the current view. Visibility will be turned on to create the corresponding tags. Do you want to continue?
>
> [Yes] [No]

6. Room tags are placed in all rooms.

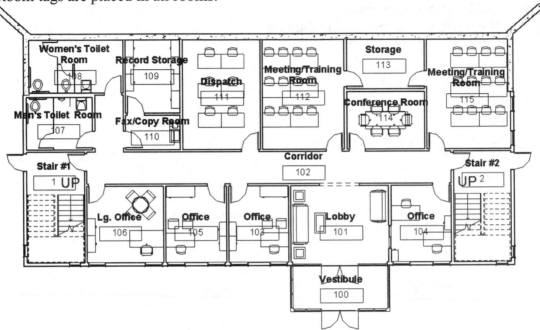

7. Highlight the view in the browser. Right click and select **Duplicate View→Duplicate**.

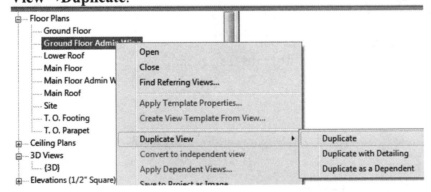

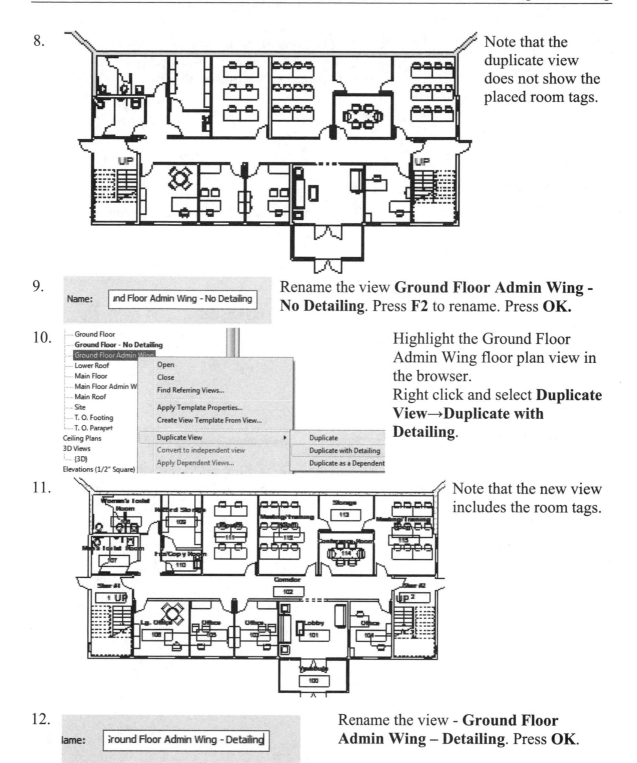

8. Note that the duplicate view does not show the placed room tags.

9. Rename the view **Ground Floor Admin Wing - No Detailing**. Press **F2** to rename. Press **OK.**

Name: ınd Floor Admin Wing - No Detailing

10. Highlight the Ground Floor Admin Wing floor plan view in the browser.
Right click and select **Duplicate View→Duplicate with Detailing**.

11. Note that the new view includes the room tags.

12. Rename the view - **Ground Floor Admin Wing – Detailing**. Press **OK**.

ıame: ;round Floor Admin Wing - Detailing

13.  Select one of the room tags. Right click and select **Select All Instances→Visible in View**.

14.  In the Type drop-down on the Properties pane, select **Room Tag with Area**.

15. The room tags update.

16. Switch back to the original Ground Floor Admin Wing. Note that the room tags for that view did not update.

17. Close without saving.

Exercise 7-5
Creating a Detail View

Drawing Name: **i_detail.rvt**
Estimated Time to Completion: 45 Minutes

Scope
Add detail components to a detail view.
Set a detail view to independent.
Change the clip offset of a detail view.
Use Repeating Details.
Use detail lines.

Solution

1. Views (all)
    Floor Plans
    — Ground Floor
    — Lower Roof
    — Main Floor
    — **Main Roof**
    — Site
    — T. O. Footing
    — T. O. Parapet

    Activate the **Main Roof** floor plan.

2.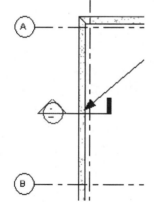

    Double click on the section head located between A-B grids to activate that section view.

3. — Callout of Section 2
    — Sections (Wall Section)
    — **Section 2**
    — Section 4

    Note the name of the active section view in the browser.

4.

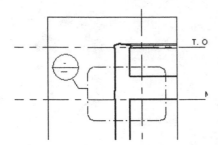

Double click on the callout bubble to activate the callout view.

5.

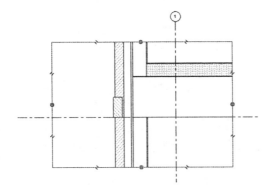

Note the name of the active callout section view in the browser.

6.

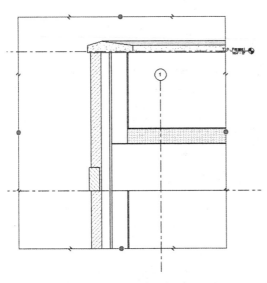

Click on the crop window to activate the grips. The grips allow you to change the extents of the view.

7.

Drag the crop region up using the grip bubble so that the roof parapet is shown.

8. 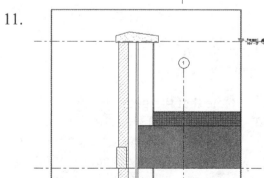 Highlight the active Callout view in the browser.
Right click and select **Rename**.

9. Change the name to **Roof Detail**.
Press **OK**.

Name: Roof Detail

10. 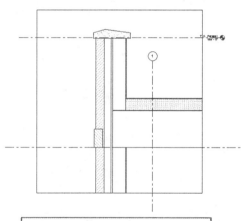 In the Properties pane:
Under Extents:
Set the Far Clip Settings to **Independent**.
Change the Far Clip Offset to **5' 0"**.

Note that the view updates and no longer displays the far North wall.

11. Select the roof.

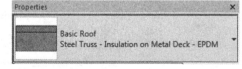

Basic Roof
Steel Truss - Insulation on Metal Deck - EPDM

If the roof is selected, it should be listed in the Properties pane.

12. 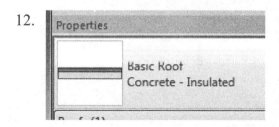 Change the Roof type to **Concrete - Insulated** on the Properties pane.

If the roof is changed in the section view, is it changed in the model?

13. Select the **Repeating Detail** tool from the Detail panel on the Annotate ribbon.

14. Set the repeating detail as **Brick** using the Type Selector.

Repeating Detail Brick

15. 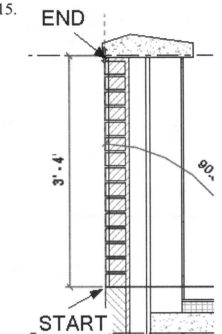 Pick the two points indicated to place the brick detail.
Start at the bottom and end at the top where the T.O. Parapet level is.

16. **Edit Type** Select the brick repeating detail just placed.
Select the **Edit Type** on the Properties pane.

17.
Parameter	
Pattern	
Detail	Brick Standard : Running Section
Layout	Fixed Distance
Inside	☑
Spacing	0' 2 5/8"
Detail Rotation	None

Note that the detail was placed using a **Fixed Distance** layout.
Note that the detail can be rotated.
Note the spacing is set to **2 5/8"**.

18. 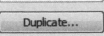 Select **Duplicate**.

19. Name: Mortar|

    Type **Mortar**.
    Press **OK**.

20.
Pattern	
Detail	Mortar Joint : Brick Joint
Layout	Fixed Distance
Inside	☑
Spacing	0' 2 5/8"
Detail Rotation	None

    For the Detail field, select **Mortar Joint: Brick Joint**. Press **OK**.
    If you accidently change the brick repeating detail to the new repeating detail, change it back using the type selector.

21. Component

    Select the **Repeating Detail** tool from the Detail panel on the Annotate ribbon.

22. Repeating Detail Mortar

    Set the repeating detail as **Mortar** using the Type Selector.

23.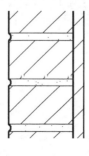

    Start on top of the first brick placed and end at the TO Parapet level.

24.

    Zoom in to see the mortar joints.

25.

    Select the **Detail Component** tool from the Annotate ribbon.

26.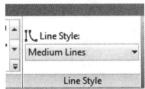

    Select the **3″ x 3″ Cant Strip** from the drop-down list on the Type Selector.

27.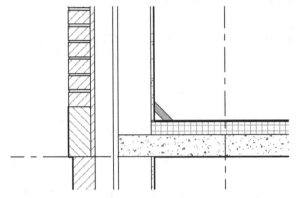

    Place the cant strip as shown.

28.

    Select the **Detail Line** tool from the Annotate ribbon.

29.

    Select **Medium Lines** from the Line Style options drop-down list.

30.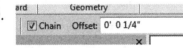

    Enable **Chain**. Enter **1/4″** in the Offset box.

31.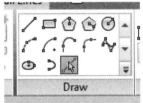

    Select **Pick Lines** mode from the Draw panel.

32.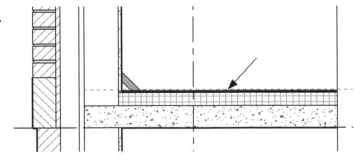

    Pick the top of the roof to place the first offset line.

33. 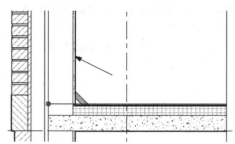 Pick the vertical wall.

34. 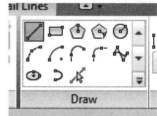 Switch to **Draw Line** mode.

35. 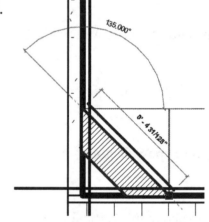 Place a line parallel to the Cant Strip.
Set the offset to 1/4" and start the line at the top of the cant strip. This will orient the line above the can strip.

36. 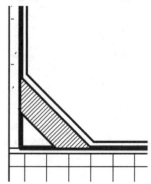 Use the **Trim** tool on the Modify Ribbon to clean up the detail lines.

37.

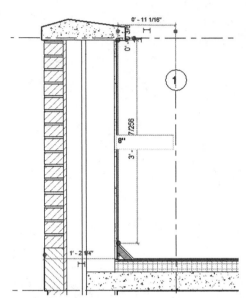

Select the vertical line to enable the temporary dimension.
Select the vertical line's vertical dimension.
Change the value to 8".

38.

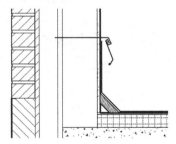

Select **Component→Detail Component** on the Annotate ribbon.

39.

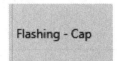
Flashing - Cap

In the Properties pane:
Select **Flashing - Cap** using the Type Selector.

40.

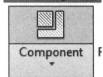

Place the Flashing Cap above the vertical line.

41.

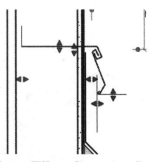

Select the flashing cap and use the grips to adjust the size.

42. Go to **File→Save As**. Save the file as *ex7-8.rvt*.

Exercise 7-6
Creating a Drafting View

Drawing Name: **i_drview.rvt**
Estimated Time to Completion: 50 Minutes

Scope
Create a drafting view.

Solution

1.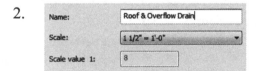
   Drafting View

   Activate the **View** ribbon.
   Select the **Drafting View** tool from the Create panel.

2. | Name: | Roof & Overflow Drain |
   | Scale: | 1 1/2" = 1'-0" |
   | Scale value 1: | 8 |

   Enter **Roof & Overflow Drain** for the Name.
   Set the Scale to **1 ½" = 1'0"**.
   Press **OK**.

3. ⌐····· Section 4
   ⊟··· Drafting Views (Detail View 1)
   ⌐····· **Roof & Overflow Drain**

   A blank view opens. Note in the browser that we are in an active drafting view.

4. Ref Plane

   Select the **Reference Plane** tool from the Work Plane panel on the Architecture ribbon.

5. Place a vertical reference plane in the middle of the blank view.

6. 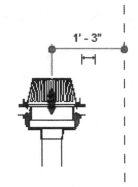   Select the **Detail Component** tool from the Detail panel on the Annotate ribbon.

Component

7.   Roof Drain   Select **Roof Drain** from the Type Selector list on the Properties pane.

8.   Place the roof drain **1′-3″** to the left of the reference plane.
Right click and select Cancel to exit the command.

1′ - 3″

9.   Select the **Filled Region** tool from the Detail panel on the Annotate ribbon.

Region

10.   Use the **Rectangle** tool to draw a rectangle as shown.

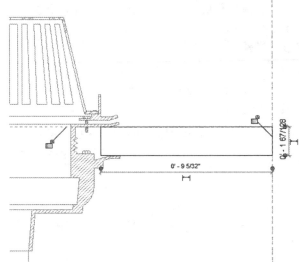

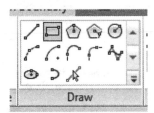

Draw

0′ - 9 5/32″

11.   Filled region
Ortho Crosshatch - Small   In the Properties pane:
Select the Filled Region type to **Ortho Crosshatch – Small**.

12.   Select the **Green Check** under Mode to **Finish Region**.

Mode

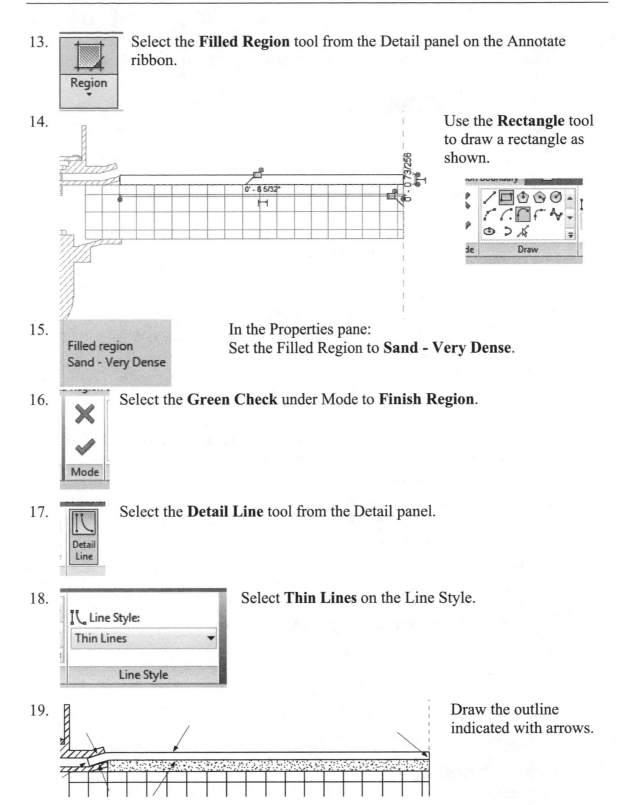

13. Select the **Filled Region** tool from the Detail panel on the Annotate ribbon.

14. Use the **Rectangle** tool to draw a rectangle as shown.

15. In the Properties pane:
Set the Filled Region to **Sand - Very Dense**.

Filled region
Sand - Very Dense

16. Select the **Green Check** under Mode to **Finish Region**.

17. Select the **Detail Line** tool from the Detail panel.

18. Select **Thin Lines** on the Line Style.

Line Style:
Thin Lines

19. Draw the outline indicated with arrows.

20. 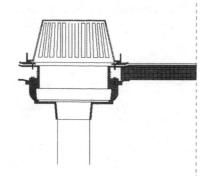   Window around the elements to select everything except the reference plane.

21. 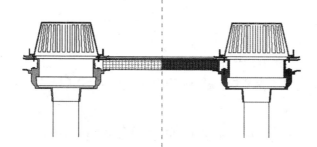   Select the **Mirror** tool using the **Mirror Pick Axis** option. Select the reference plane to use as the axis.

22. 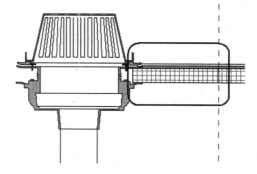   Window around the filled regions and detail lines to select.

23. 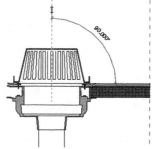   Select the **Mirror** tool using the **Draw Mirror Axis** option.

24.   Select the midpoint of the roof drain to draw the axis.

25. 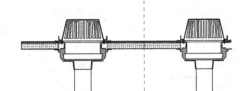 The elements are mirrored to the other side.

26. 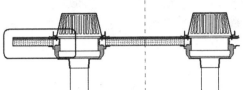 Select the mirrored elements.

27.  Select **Mirror → Pick Mirror Axis** from the ribbon.

28. 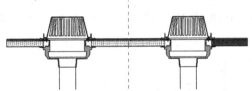 Select the reference plane to use as the mirror axis.
The elements are mirrored.

29. 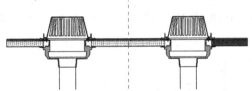 Select the **Detail Component** tool from the Annotate ribbon.

Component

30. Break Line — Select **Break Line** from the Type Selector drop-down list on the Properties pane.

31. Position the cursor to the left of the detail view.
Press the SPACEBAR to rotate the break line.

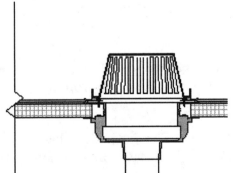

32.

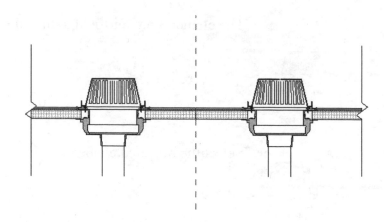

Place the break line so it is coincident with the vertical line at the end of the filled region element.
Repeat for the right side.

33. Select the left roof drain.
34. On the Properties Pane: Change the Ring Height to **2″**.

35.

Select the **Aligned** dimension tool on the Annotate ribbon.

36. Place a dimension from center to center distance between the roof drains.

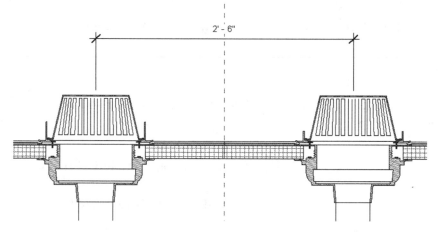

37.

Linear Dimension Style
Linear w Center - 3/32″ Arial

With the dimension selected, select **Linear w Center - 3/32″ Arial** from the Type Selector drop-down list on the Properties pane.

38.

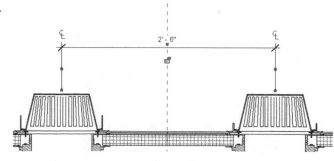

The dimension changes to display centerline symbols.

39. 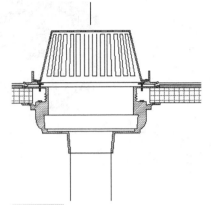 Zoom into the right roof drain.

40. Select the **Text** tool from the Annotate ribbon.

41. 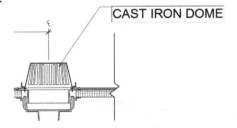 Enable the **Two Segments** leader option on the Format panel.

42. Place the text as shown.

CAST IRON DOME

43. Add the text shown.

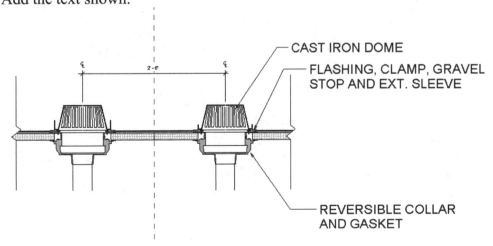

CAST IRON DOME

FLASHING, CLAMP, GRAVEL
STOP AND EXT. SLEEVE

REVERSIBLE COLLAR
AND GASKET

44. Close without saving.

Exercise 7-7
Callout from Sketch

Drawing Name: **callout_sketch.rvt**
Estimated Time to Completion: 25 Minutes

Scope
Object Settings
Callout by Sketch
Drafting View
Duplicate view as Dependent
Visibility/Graphics
Masking Region
Convert Dependent view to Independent

Solution

1. Activate the Manage ribbon.
   Select **Object Styles**.

   ; Object
   Styles

2.

Category	Line Weight Projection	Line Color	Line Pattern
Adaptive Points	1	Black	Solid
Area Tags	1	Black	Solid
Assembly Tags	1	Black	
Brace in Plan View Symbols	2	Black	Hidden 1/8"
Callout Boundary	1	Blue	Dash dot dot
Callout Leader Line	1	Black	Solid
Callout Heads	1	Black	Solid

   Model Objects · Annotation Objects · Analytical Model Objects · Imported Objects
   Filter list: Architecture

   Select the Annotation tab.
   Set the color of the Callout Boundary to **Blue**. Set the Line Pattern to **Dash dot dot.**
   Press **OK**.

3. Floor Plans
   Level 1
   Level 2
   Presentation- Level 1
   Roof
   Site

   Activate **Level 1**.

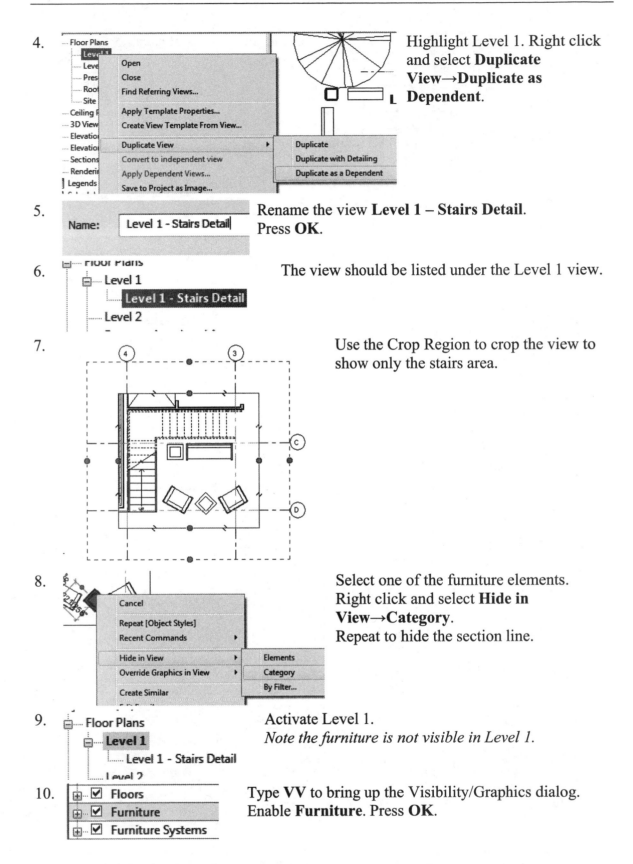

4. Highlight Level 1. Right click and select **Duplicate View→Duplicate as Dependent**.

5. Rename the view **Level 1 – Stairs Detail**. Press **OK**.

6. The view should be listed under the Level 1 view.

7. Use the Crop Region to crop the view to show only the stairs area.

8. Select one of the furniture elements. Right click and select **Hide in View→Category**. Repeat to hide the section line.

9. Activate Level 1. *Note the furniture is not visible in Level 1.*

10. Type **VV** to bring up the Visibility/Graphics dialog. Enable **Furniture**. Press **OK**.

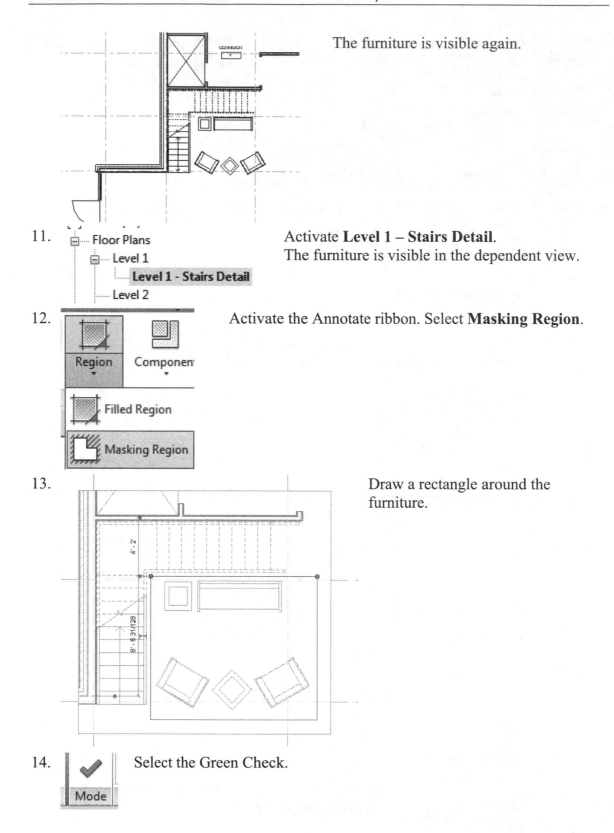

The furniture is visible again.

11. Floor Plans
    └ Level 1
        └ **Level 1 - Stairs Detail**
    └ Level 2

Activate **Level 1 – Stairs Detail**.
The furniture is visible in the dependent view.

12. Region | Componen
    Filled Region
    Masking Region

Activate the Annotate ribbon. Select **Masking Region**.

13.

Draw a rectangle around the furniture.

14. ✓ Mode

Select the Green Check.

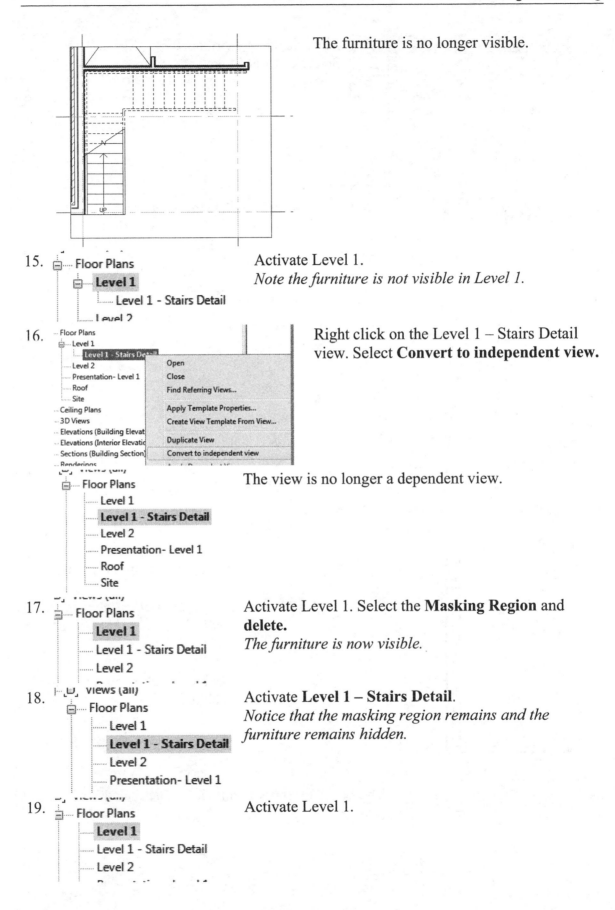

The furniture is no longer visible.

15. Activate Level 1.
Note the furniture is not visible in Level 1.

16. Right click on the Level 1 – Stairs Detail view. Select **Convert to independent view.**

The view is no longer a dependent view.

17. Activate Level 1. Select the **Masking Region** and **delete.**
The furniture is now visible.

18. Activate **Level 1 – Stairs Detail**.
Notice that the masking region remains and the furniture remains hidden.

19. Activate Level 1.

20. Activate the View ribbon.
Select **Callout Sketch**.

21. On the Options bar:
Enable **Reference other view**.
Select **Level 1 – Stairs Detail**.

22. Use the LINE tool to draw an outline around the stairs.

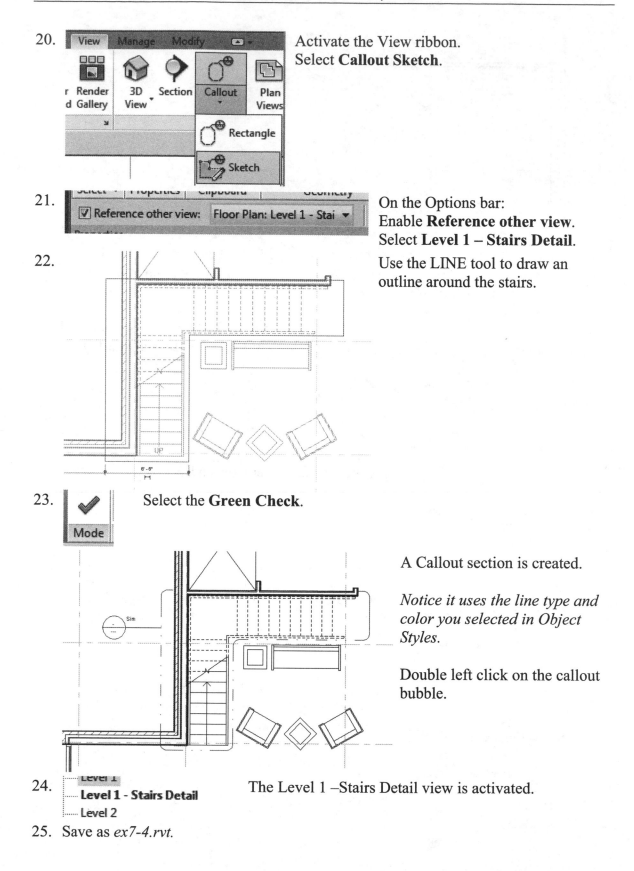

23. Select the **Green Check**.

A Callout section is created.

Notice it uses the line type and color you selected in Object Styles.

Double left click on the callout bubble.

24. Level 1
Level 1 - Stairs Detail
Level 2

The Level 1 –Stairs Detail view is activated.

25. Save as *ex7-4.rvt*.

Exercise 7-8
Save and Re-Use a Drafting View

Drawing Name: **c_ details.rvt**
Estimated Time to Completion: 10 Minutes

Scope
Save a drafting view for re-use.
Insert the saved file into a project.

Solution

1. 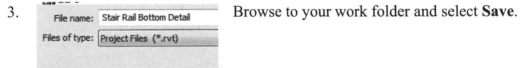 Highlight the **Stair Rail Bottom Detail** in the Drafting Views (Detail) category in the Project Browser.

2. Right click and select **Save to New File**.

3. 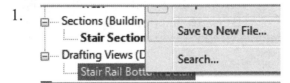 Browse to your work folder and select **Save**.

4. Start a new project.

5. Select the **Architectural Template**.
   Press **OK**.

6.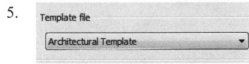

   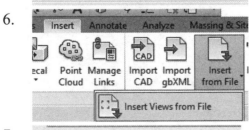

   Activate the **Insert** ribbon.
   Select **Insert from File→Insert Views from File** from the Import panel.

7. Locate the file you just saved.
   Press **Open**.

8.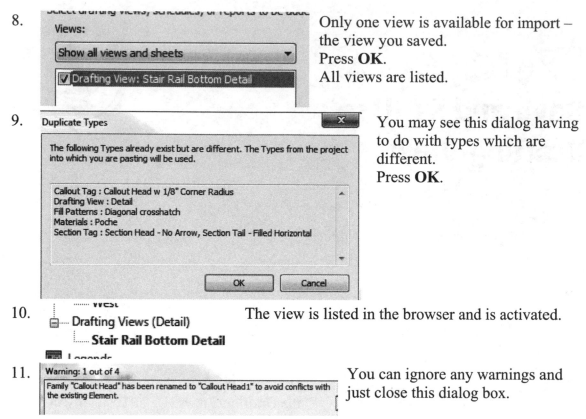

Only one view is available for import – the view you saved.
Press **OK**.
All views are listed.

9.

You may see this dialog having to do with types which are different.
Press **OK**.

10. The view is listed in the browser and is activated.

11. You can ignore any warnings and just close this dialog box.

12. Close the files without saving.

Exercise 7-9
Calculated Values in Tags and Schedules

Drawing Name: **occupancy factor.rvt**
Estimated Time to Completion: 30 Minutes

Scope
Create a schedule using a calculated value
Create a room tag using a calculated value
Modify a room size

Solution

1. [Views (all) / Floor Plans / Ground Floor / **Ground Floor Admin Wing** / Lower Roof / Main Floor / Main Floor Admin Wing / Main Roof / Site]

   Activate the **Ground Floor Admin Wing** floor plan.

2.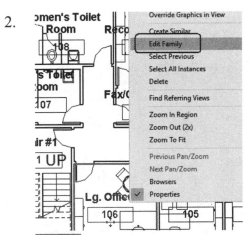

   Select one of the room tags in the view.
   Right click and select **Edit Family**.

3. A Label

   On the Create ribbon:
   Select the **Label** tool from the ribbon.

4. Select **Add calculated parameter**.

5.

Calculated Value ✕

Name:	Max Occupancy
Discipline:	Common ⌄
Type:	Number ⌄
Formula:	rounddown((Area/1 SF)/0.3) ...

| OK | Cancel | Help |

Type **Max Occupancy**.
Set the Type of Parameter as **Number**.
For the formula:
Type:
rounddown((Area/1 SF)/0.3)
0.3 is the occupancy factor.
Press **OK**.

6. onment.   ☐ Wrap between parameters only

Label Parameters

	Parameter Name	Spaces	Prefix	Sample Value	Suffix
1	Max Occupancy	1	Max OCC =	Max Occupancy	

In the Prefix column:
Type **Max OCC =**
Press **OK**.

7.

Room name

101

V50uSE
Max OCC = Max
Occupancy

Position the label below the area.

8.

File name:	Room Tag with Occupancy.rfa
Files of type:	Family Files (*.rfa)

Save the family as *Room Tag with Occupancy.rfa* in your work folder.

9. Select **Load into Project**.

Load into
Project

10.

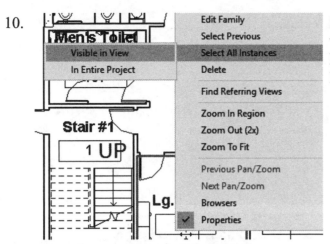

Select one of the room tags in the view.
Right click and select **Select all Instances→Visible in View.**

11.

Locate the **Room Tag with Occupancy** in the Type Selector.
Select the **Room Tag with Area**.

12.

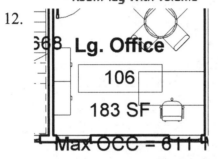

The room tags update.

13.

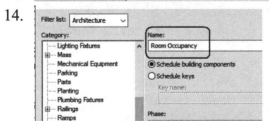

Activate the View ribbon.
Select **Schedules→Schedule/Quantities.**

14.

Highlight **Rooms**.
Change the Name to **Room Occupancy**.
Press **OK**.

15.

Scheduled fields (in order):

Number
Name
Area

Add the fields:
- Number
- Name
- Area

16.   Select **Add Calculated Value**.

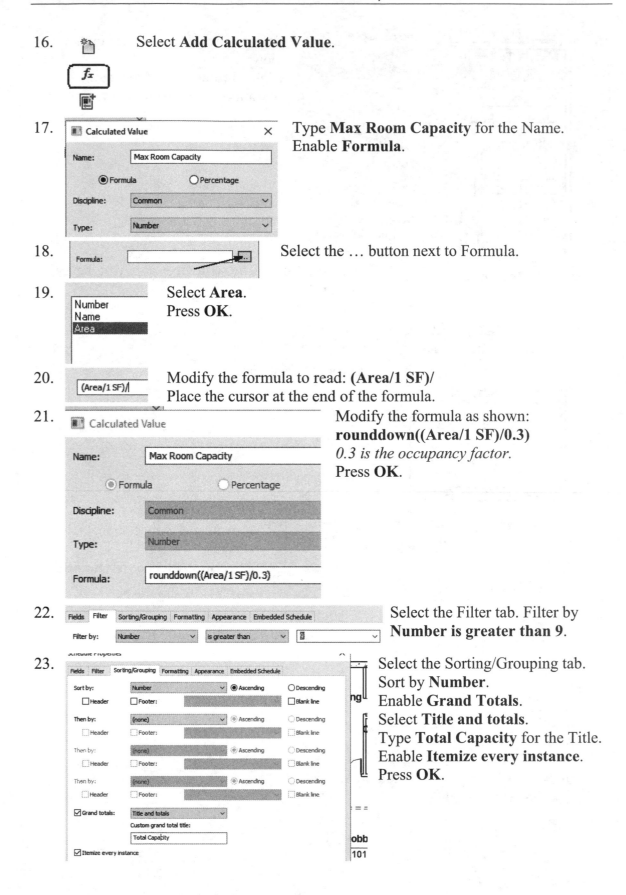

17.   Type **Max Room Capacity** for the Name.
Enable **Formula**.

18.   Select the … button next to Formula.

19.   Select **Area**.
Press **OK**.

20.   Modify the formula to read: **(Area/1 SF)/**
Place the cursor at the end of the formula.

21.   Modify the formula as shown:
rounddown((Area/1 SF)/0.3)
0.3 is the occupancy factor.
Press **OK**.

22.   Select the Filter tab. Filter by
Number is greater than 9.

23.   Select the Sorting/Grouping tab.
Sort by **Number**.
Enable **Grand Totals**.
Select **Title and totals**.
Type **Total Capacity** for the Title.
Enable **Itemize every instance**.
Press **OK**.

24.

<Room Occupancy>			
A	B	C	D
Number	Name	Area	Max Room Capacity
100	Vestibule	136 SF	454
101	Lobby	241 SF	804
102	Corridor	410 SF	1368
103	Office	143 SF	476
104	Office	143 SF	476
105	Office	143 SF	476
106	Lg. Office	183 SF	611
107	Men's Toilet Room	108 SF	359
108	Women's Toilet Ro	155 SF	517
109	Record Storage	123 SF	410
110	Fax/Copy Room	58 SF	192
111	Dispatch	262 SF	872

The schedule appears.

25.

Activate the **Ground Floor Admin Wing** floor plan.

26.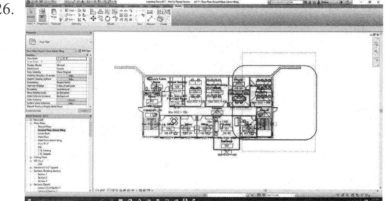

Use a crossing to select the right side of the wing.

27. Select the **MOVE** tool.

28. ☑ Constrain ☐ Disjoin ☐ Multiple

Enable **Constrain** on the Options bar.

29.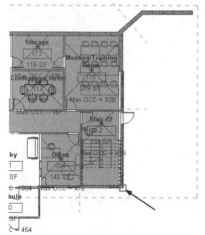

Select the lower right corner as the basepoint.

30.

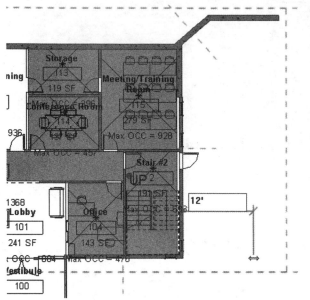

Move the cursor to the right.
Type **12'**.
Press ENTER.

31.

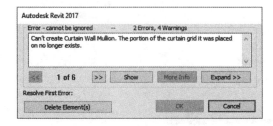

If you see this error dialog, press **Delete Element(s)**.

32.

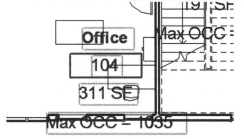

Note that the room tags updated on the affected rooms.

33. Save as **ex7-7.rvt**.

Exercise 7-10
Creating a Detail Group

Drawing Name: **i_detail_groups.rvt**
Estimated Time to Completion: 20 Minutes

Scope
Add hidden lines.
Add a steel plate using Detail Component
Add bolts to a plate.
Create a detail group.
Mirror a detail group.

Solution

1. Elevations (Building Elevation)
   Sections (Wall Section)
   Beam-Column Connection
   Section 1
   Section 2
   Section 3

   Activate the **Beam-Column Connection** section view.

2. Component

   Activate the Annotate ribbon.
   Select the **Detail Component** tool from the Detail panel.

3. Load Family
   Mode

   Select **Load Family** from the Mode panel.

4. ProgramData
   Autodesk
   RVT 2017
   Libraries
   US Imperial
   Detail Items
   Div 05-Metals
   050500-Common Work Results for Metals
   050523-Metal Fastenings

   Browse to the *Metal Fastenings* folder under *Detail Items\ Div 05 - Metals\ 050500 - Common Work Results for Metals*.

5. File name: Steel Shear Plate-Face
   Files of type: All Supported Files (*.rfa, *.adsk)

   Open the *Steel Shear Plate - Face* family.

6.

Press the SPACE bar to rotate the component prior to placing.
Place on the left side of the column.

7.

Activate the Annotate ribbon.
Select the **Repeating Detail Component** tool from the Detail panel.

8.

Repeating Detail
A325 - 5/8" Plan

Select the **A325 - 5/8″** Plan detail on the Properties pane.

9.

Edit Type

Select **Edit Type** in the Properties pane.

10.

Pattern	
Detail	Bolt-High-Strength-Plan : A325 - 5/8"
Layout	Fixed Number
Inside	
Spacing	0' 3"
Detail Rotation	None

Set the Layout to **Fixed Number**.
Press **OK**.

11.

Dimensions	
Number	5
Length	1' 1 1/2"
Identity Data	

In the Properties pane:
Change the number to **5**. Press **Apply**.

12.

Place a bolt in the bottom hole. Then pick the top hole. Right click and select Cancel to exit the command.
*If the number reverts back to 3, place the detail. Release it. Re-select it and then change it again in the Properties pane and press **Apply**.*

13. Window around the steel plate and the placed bolts to select them. Select **Detail Group→Create Group** from the Detail panel on the Annotate ribbon.

14. Type **Plate and Bolts** for the Detail Group name. Press **OK**.

Name: Plate and Bolts

☐ Open in Group Editor

OK   Cancel

15. In the Browser, locate the detail group created under **Groups→Detail**.

Groups
  Detail
    [A] Plate and 5 bolts in section
    [A] Plate and Bolts

16. Select the steel plate and bolts detail group in the display window. The group should highlight. Select **Mirror→Pick Axis** from the Modify panel.

17. Select the grid centered on the column as the axis. Click anywhere in the window to release the selection and complete the mirror command.

18. Locate the **Plate and 5 bolts in section** in the Project Browser.

Families
Groups
  Detail
    [A] Plate and 5 bolts in section
    [A] Plate and Bolts

19. Drag and drop them into the location shown – at the center. Left click to place.

20. Close without saving.

Exercise 7-11
Revision Control

Drawing Name: **i_Revisions.rvt**
Estimated Time to Completion: 15 Minutes

Scope
Add a sheet.
Add a view to a sheet.
Setting up Revision Control in a project.

Solution

1. Activate the **View** ribbon. Select the **New Sheet** tool on the Sheet
    Composition panel.

    Sheet

2. Load...   Select the **Load** button.

3. Titleblocks

    Name
    A 8.5 x 11 Vertical
    B 11 x 17 Horizontal
    C 17 x 22 Horizontal
    D 22 x 34 Horizontal
    E 34 x 44 Horizontal
    E1 30 x 42 Horizontal

    Locate the *Titleblocks* folder.
    Locate the **D 22 x 34 Horizontal** titleblock.
    Press **Open**.

4. Select titleblocks:

    D 22 x 34 Horizontal
    E1 30 x 42 Horizontal : E1 30x42 Horizontal
    None

    Highlight the **D 22 x 34 Horizontal** titleblock.
    Press **OK**.

5.

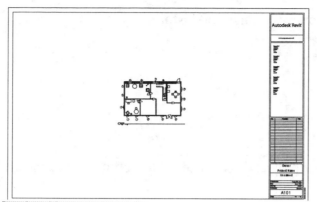

Drag and drop the **Level 1** floor plan onto the sheet.

6.

Graphics	
View Scale	1/4" = 1'-0"
Scale Value 1:	48

Select the view.
In the Properties pane:
Set the View Scale to **1/4″ = 1′-0″**.

7.

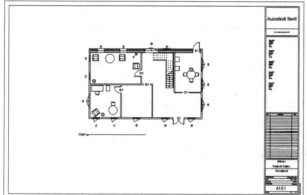

Position the view on the sheet.

8.

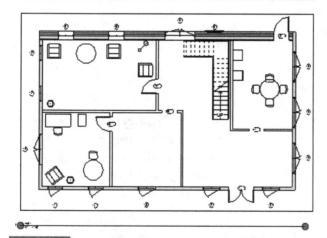

To adjust the view title bar, select the view and the grips will become activated.

9.

Revisions (

Activate the **View** ribbon. Select **Revisions** on the **Sheet Composition** panel.

10. This dialog manages revision control settings and history.

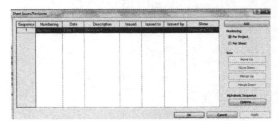

Numbering can be controlled per project or per sheet. The setting used depends on your company's standards.

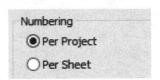

Enable **Per Project**.
One Revision is available by default. Additional revisions are added using the **Add** button.

11. 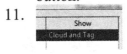 The visibility of revisions can be set to **None**, **Tag** or **Cloud and Tag**. Use the **None** setting for older revisions which are no longer applicable so as not to confuse the contractors.
Set the revision to show **Cloud and Tag**.

12. Select **Alphanumeric** under options.

13. Change the sequence to remove the letters **I** and **O**.

14. Select the **Numeric** tab.
Type **A** for the prefix.
This would allow revisions to be created using A1, A2, A3…
Press **OK**.

15. Enable the **Numeric** option. The dialog will pop up to show the current settings for Numeric revisions.
Press **OK**.

16. Enter the three revision changes shown.

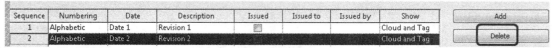

Sequence	Revision Number	Numbering	Date	Description	Issued	Issued to	Issued by	Show
1	A1	Numeric	08.02	Wall Finish Change	☐	Joe	Sam	Cloud and Tag
2	A2	Numeric	08.04	Window Style Change	☐	Joe	Sam	Cloud and Tag
3	A3	Numeric	08.13	Door Hardware Change	☐	Joe Sam		Cloud and Tag

17. Press **OK** to close the dialog.
    You can delete revisions if you make a mistake. Just highlight the row and press
    Delete.

Sequence	Numbering	Date	Description	Issued	Issued to	Issued by	Show		
1	Alphabetic	Date 1	Revision 1	☑			Cloud and Tag	Add	
2	Alphabetic	Date 2	Revision 2				Cloud and Tag	Delete	

18. Save the project as *ex7-11.rvt*.

Exercise 7-12

Modify a Revision Schedule

Drawing Name: **ex7-11.rvt**
Estimated Time to Completion: 20 Minutes

Scope
Modify a revision schedule in a title block.

Solution

1.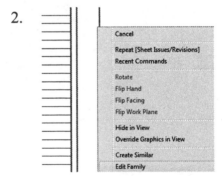
    Zoom in to the **Revision Block** area on the sheet. Note that the title block includes a revision schedule by default.

2.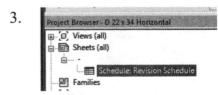
    Select the title block.
    Right click and select **Edit Family**.

3. Select the **Revision Schedule** in the Project Browser.

4. 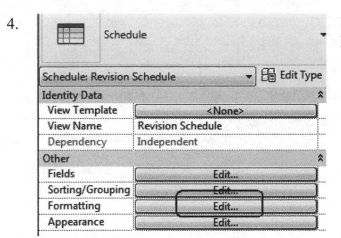 Select **Edit** next to Formatting in the Properties pane.

5.  Change the First Column Header to **Rev**.

6. Select the Fields tab.
Add the **Issued By** field.

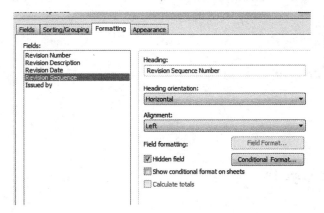 Do **NOT** remove the Revision Sequence field. This is a hidden field.

Note that the Hidden field control is located on the Formatting tab. This is a possible question on the certification exam.

7. 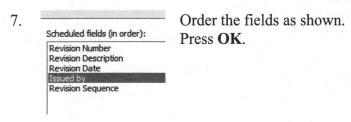 Order the fields as shown.
Press **OK**.

8. The Revision Schedule updates.

9. ⊞ Schedules
   ⊟ Sheets (all)
       ⊟ -
           ▦ Schedule: Revision Schedule

   To return to the title block sheet view, double left click on the – below Sheets (all) in the browser.

10.

   Adjust the column width of the schedule so it fits properly in the title block.

Rev	Description	Date	Issued by

   Client Name

   Project Name

   Sheet Name

11. File name: D rev schedule

   Files of type: Family Files (*.rfa)

   Go to **File→SaveAs→Family**.
   Save as *D rev schedule.rfa*.

12. Load into Project

   Family Editor

   Select **Load into Project** on the Family Editor panel on the ribbon.

13. If you have more than one project open:
   Place a check next to *ex7-11.rvt*.
   Press **OK**.

14.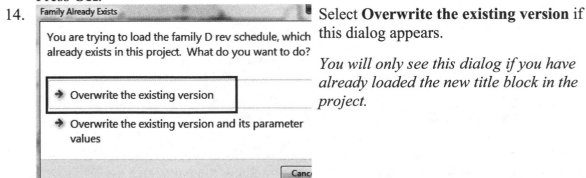

   Family Already Exists

   You are trying to load the family D rev schedule, which already exists in this project. What do you want to do?

   ➡ Overwrite the existing version

   ➡ Overwrite the existing version and its parameter values

   Cance

   Select **Overwrite the existing version** if this dialog appears.

   You will only see this dialog if you have already loaded the new title block in the project.

15.

Warning

Can't create this kind of element in this view in the current mode.

If you see this dialog, just close it. This dialog will appear in a floor plan view.

16.

Sheets (all)
  A100 - Cover Sheet
  A101 - Unnamed
  Families

Activate the Unnamed Sheet.

17.

D rev schedule

Select the title block so it is highlighted.
Locate the **D rev schedule** title block in the pull-down list on the Properties pane and select.

18.

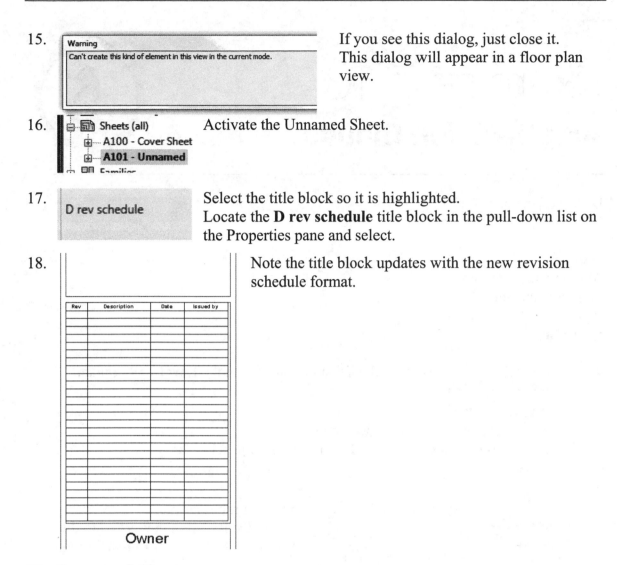

Note the title block updates with the new revision schedule format.

19. Save as *ex7-12.rvt*.

Exercise 7-13

Add Revision Clouds

Drawing Name: **ex7-12.rvt**
Estimated Time to Completion: 30 Minutes

Scope
Add revision clouds to a view.
Tag revision clouds.

Solution

1. Sheets (all)
    A100 - Cover Sheet
    A101 - Unnamed
    Families

    Activate the Unnamed sheet with the **Level 1** floor plan.

2. Activate the **Annotate** ribbon.
    Select **Revision Cloud** from the Detail panel.

    Revision
    Cloud

3. | Revision Clouds | |
    |---|---|
    | **Identity Data** | |
    | Revision | Seq. 1 - Wall Finish Change |
    | Revision Number | A1 |
    | Revision Date | 08.02 |

    On the Properties pane:
    Select the Revision that is tied to the revision cloud – **Wall Finish Change**.

4. | Revision Clouds | ⟶ Edit Type |
    |---|---|
    | **Identity Data** | ⌃ |
    | Revision | Seq. 1 - Wall Finish Change |
    | Revision Number | A1 |
    | Revision Date | 08.02 |
    | Issued to | Joe |
    | Issued by | Sam |
    | Mark | |
    | Comments | Finish Changed to SW Gold Sunset |

    Under Comments: Enter **Finish Changed to SW Gold Sunset**.

5.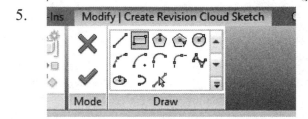

    You can use any of the available Draw tools to create your revision cloud.
    Select the **Rectangle** tool.

7-58

6. 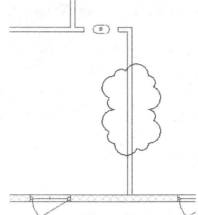 Draw the Revision Cloud on the wall indicated.

Note that a cloud is placed even though you selected the rectangle tool.

7. Select the **Green Check** on the Mode panel to **Finish Cloud**.

8. Activate the **Annotate** ribbon.
Select **Revision Cloud**.

9. 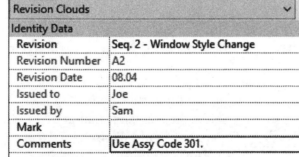 Select the Revision that is tied to the revision cloud – **Window Style Change**.

Under Comments: Enter **Use Assy Code 301**.

Revision Clouds	⌄
Identity Data	
Revision	Seq. 2 - Window Style Change
Revision Number	A2
Revision Date	08.04
Issued to	Joe
Issued by	Sam
Mark	
Comments	Use Assy Code 301.

10. 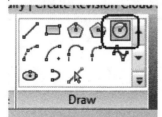 Select the **Circle** tool.

11.

Draw the Revision Cloud on the window indicated.

Note that a cloud is placed even though you selected the circle tool.

12.

Select the **Green Check** on the Mode panel to **Finish Cloud**.

13.

If you mouse over the revision cloud, you will see a tooltip to indicate what the revision is.

Revision Clouds : Revision Cloud: A2 - Window Style Change

14.

Activate the **Annotate** ribbon.
Select **Revision Cloud** from the Detail panel.

15.

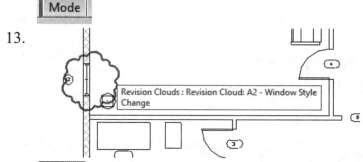

Revision Clouds (1)	
Identity Data	
Revision	Seq. 3 - Door Hardware Change
Revision Number	A3
Revision Date	08.13
Issued to	Joe
Issued by	Sam
Mark	
Comments	Use Hardware Set #230.

On the Properties pane:
Select the Revision that is tied to the revision cloud – **Door Hardware Change**.
Under Comments: Enter **Use Hardware Set #230**.

16.

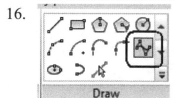

Select the **Spline** tool.

Draw the cloud in the clockwise direction.

17. 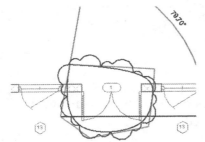 Draw the Revision Cloud on the door indicated.

18. **Error - cannot be ignored**

Self intersecting splines are not allowed in sketches.

Show

You might see this error if the spline has overlapping segments. If you get this error, delete the spline and try again. You do not have to close the spline in order to create the revision cloud.

19. vision Cloud

✗
✓
Mode

Select the **Green Check** on the Mode panel to **Finish Cloud**.

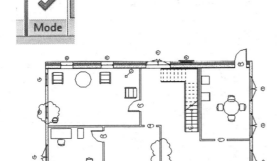 Three revision clouds have been placed in the view.

20.

Rev	Description	Date	Issued by
A1	Wall Finish Change	08.02	Sam
A2	Window Style Change	08.04	Sam
A3	Door Hardware Change	08.13	Sam

Zoom into the title block and note that the revision block has updated with the revisions that have been issued.

21. Tag by Category

Select the **Tag by Category** tool from the Tag panel on the Annotate ribbon.

22.

Revision Clouds : Revision Cloud: B - Window Style Change

Pick one of the revision clouds to identify the category to be tagged.

23. There is no tag loaded for Revision Clouds. Do you want to load one now?

If you see this message. Press **Yes**.

24.

North Arrow 2
Part Tag
Revision Tag
Room Tag
Section Head - 1 po

Select the *Annotations* folder.
Locate the **Revision Tag**.
Press **Open**.

25.

Select the revision cloud to add the tag.
The tag is placed.

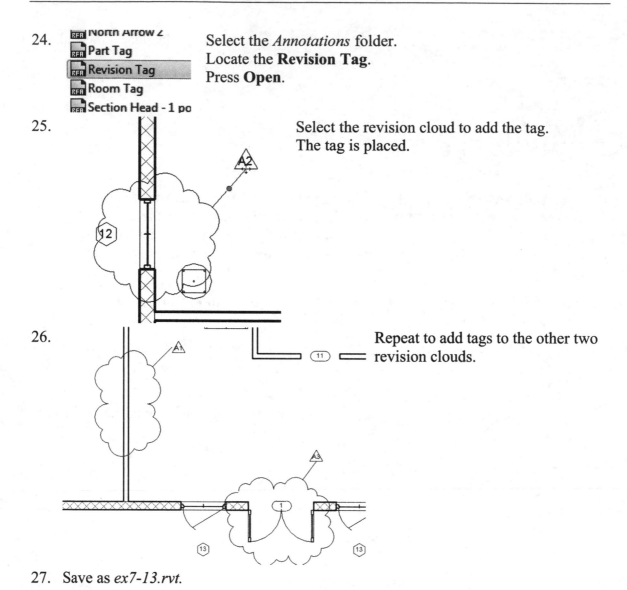

26.

Repeat to add tags to the other two revision clouds.

27. Save as *ex7-13.rvt*.

Exercise 7-14

Aligning Views between Sheets

Drawing Name: **aligning_views.rvt**
Estimated Time to Completion: 30 Minutes

Scope
Using Grid Guides

Solution

1. 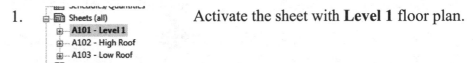   Activate the sheet with **Level 1** floor plan.

2. 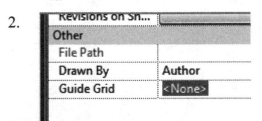   In the Properties pane:
   Scroll down and note that Grid Guide is set to
   <None>.

3.    Activate the **View** ribbon.
   Select **Guide Grid** on the Sheet Composition panel.

4.   Press **OK**.

5. | Dimensions | |
   | --- | --- |
   | Guide Spacing | 2" |
   | Identity Data | |
   | Name | Guide Grid 1 |

   Select the Guide Grid.
   Select by left clicking on an edge.
   In the Properties pane:
   Set the Guide Spacing to **2"**.
   Press **Apply**.

6.

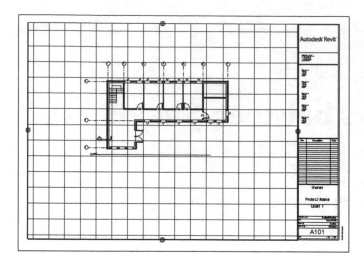

The grid updates.
Use the blue grips to adjust the grid so it lies entirely inside the title block.

7.

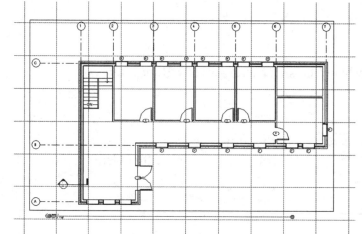

Select the viewport.

Select the **Move** tool from the Modify Panel.

8.

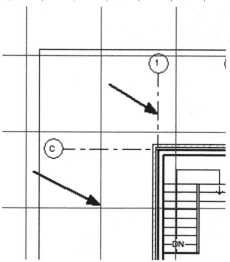

Select on the grid line 1 in the view.
Then select the guide grid.

9.

The view snaps to align with the grid.
Repeat to shift grid line C into alignment with the grid guide.

10.

Zoom out and note which grid the view is aligned to.

11.

— Schedules/Quantities
⊟ Sheets (all)
 ⊞ A101 - Level 1
 ⊞ **A102 - High Roof**
 ⊞ A103 - Low Roof

Activate the sheet named **High Roof**.

12.

Other	
File Path	
Drawn By	Author
Guide Grid	Guide Grid 1

In the Properties pane:
Set the Guide Grid to **Guide Grid 1**.

13. Select the viewport.

Select the **Move** tool from the Modify Panel.

14.

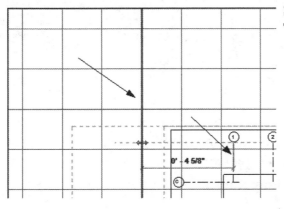

Select on the grid line 1 in the view.
Then select the guide grid.

15.

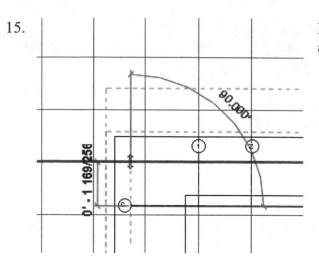

Repeat to shift grid line C into alignment with the grid guide.

16.

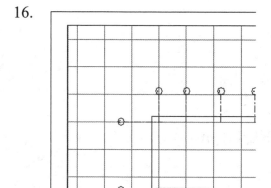

Verify that the view is aligned to the same guide grid cells as sheet A101.

17.

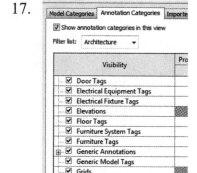

Type **VV**. Select the **Annotation Categories** tab. Disable visibility of the **Guide Grid**. Press **OK**.

18.

Select the viewport. Select **Pin** from the Modify panel.
This will lock the view to its current position.

19.

Activate the **Low Roof** sheet.

20. 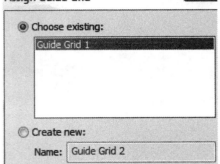 Select **Guide Grid** from the View ribbon.

21. Note you can select the existing guide grid.
Press **OK**.

Assign Guide Grid ☒

⦿ Choose existing:

Guide Grid 1

○ Create new:

Name: Guide Grid 2

22. Note in the Properties pane the Guide Grid 1 is listed.

Other	
File Path	
Drawn By	Author
Guide Grid	Guide Grid 1

23. Select the view. On the Options bar:
Set the Rotation on Sheet to **90° Clockwise**.

Rotation on Sheet: 90° Clockwise ▼

24. 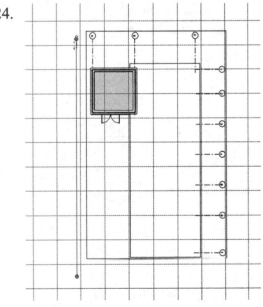 The view rotates.

25. Close the file without saving.

Certified User Practice Exam

1. Select the element which is NOT considered an annotation:

   A. Revision Cloud
   B. Elevation Marker
   C. Aligned Dimension
   D. Door Tag
   E. Text

2. Tags:

   A. Are View-specific
   B. Cannot be placed in an elevation view
   C. Are added or modified using the Text tool
   D. Display Parametric Information
   E. Scale with a view

3. A callout:

   A. Crops a 3D view
   B. Tags Elements in a building model
   C. Shows an enlarged version of part of the parent view
   D. Generates separate views of elements of existing views

4. A detail view can be placed in which of the following?

   A. 3D View
   B. Plan View
   C. Elevation View
   D. Section View
   E. Camera View

5. Drafting Views can contain (select all that apply)?

   A. Model elements
   B. Detail Lines
   C. Detail Components
   D. Filled Region
   E. Model Lines

6. A view title on a sheet can be moved independently from the view portion of the viewport, if you:

   A. Select only the view title.
   B. Enable Move View Title Independently on the Properties pane.
   C. Disable Move Title with View on the Properties pane.
   D. Modify the Viewport's Type parameters.

7. You can place Detail Components in all views EXCEPT:

    A. Plan
    B. Sections
    C. Sheets
    D. Legends

8. A Filled Region allows you to define its:

    A. Function
    B. Level Offset
    C. Boundary
    D. Phase

Answers
    1) B; 2) A; 3) C; 4) B, C & D; 5) B, C & D; 6) A; 7) C; 8) C;

Certified Professional Practice Exam

1. _____ can be used to align elements within and between sheets.

    A. Guide Grid
    B. Align
    C. Underlay
    D. Aligned Dimension

2. To lock a view's position on a sheet:

    A. Right click and select Lock Position.
    B. Select the viewport, then Enable Lock Position on the Properties pane.
    C. Change the viewport's properties to lock position.
    D. Select the viewport, then use Pin from the Modify panel to lock position.

3. **True or False**: Revisions can include letters (alpha) and numbers.

4. **True or False**: After you place a revision cloud, you can add a tag to it.

5. **True or False**: A revision schedule displays information derived from revision clouds.

6. To rotate a view on a sheet:

    A. Select the view and then use the Rotate tool on the Modify panel.
    B. Set the view rotation in the Properties pane.
    C. Select the view and then set the view rotation on the Options bar.
    D. Change the Project North.

Answers
1) B; 2) D; 3) True; 4) True; 5) True; 6) A

Construction Documentation

This lesson addresses the following certification exam questions:

- Schedules
- Legends
- Rooms and Areas

A schedule is a type of view. It is a tabular display of information using element properties. Each element property is represented as a column field in the schedule.

The Schedules and Legends tools are located on the View ribbon.

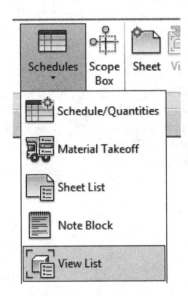

The **Schedule/Quantities** tool is used to create component schedules. Components are elements such as doors, windows, and rooms.

The **Material Takeoff** schedule lists the materials of any family used in the project.

A **Sheet List** is usually placed on the first sheet of the documentation set. It is a schedule of all sheets used in the project.

Note Blocks are useful for listing notes applied to elements in a project. For example, a user may have attached a note to several walls, which might have a description of the finish applied to each wall.

A **View List** is used to sort and itemize the views available in a project.

Exercise 8-1
Creating a Door Schedule

Drawing Name: **i_schedules.rvt**
Estimated Time to Completion: 5 Minutes

Scope
Create a door schedule.
Use the sort and group feature to determine how many doors of a specific type are on a level.

Solution

1.
Activate the **View** tab on the ribbon.
Select **Schedule/Quantities** from the Create panel.

2.
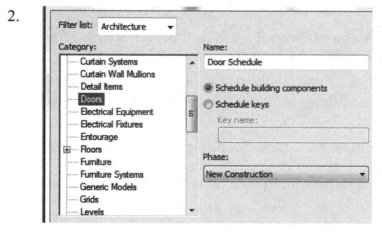
Highlight the **Doors** category.
Enable **Schedule building components**.
Set the Phase to **New Construction**.
Press **OK**.

3.

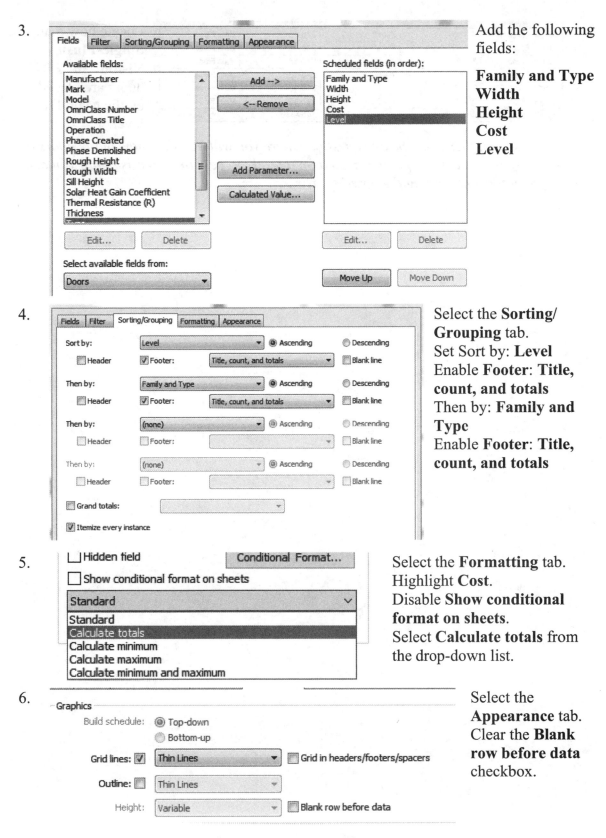

Add the following fields:

Family and Type
Width
Height
Cost
Level

4.

Select the **Sorting/ Grouping** tab.
Set Sort by: **Level**
Enable **Footer**: **Title, count, and totals**
Then by: **Family and Type**
Enable **Footer**: **Title, count, and totals**

5.

Select the **Formatting** tab.
Highlight **Cost**.
Disable **Show conditional format on sheets**.
Select **Calculate totals** from the drop-down list.

6.

Select the **Appearance** tab.
Clear the **Blank row before data** checkbox.

7. Press **OK**.

8.

Single-Flush Vi	3' - 0"	7' - 0"	240.15	Main Floor
Single-Flush Vi	3' - 0"	7' - 0"	240.15	Main Floor
Single-Flush Vi	3' - 0"	7' - 0"	240.15	Main Floor
Single-Flush Vi	3' - 0"	7' - 0"	240.15	Main Floor
Single-Flush Vi	3' - 0"	7' - 0"	240.15	Main Floor
Single-Flush Vi	3' - 0"	7' - 0"	240.15	Main Floor
Single-Flush Vision: 36" x 84": 6			1440.90	

Locate how many **Single-Flush Vision: 36" x 84"** doors are placed on the Main Floor level.

On the exam, there may be one question where you will be asked how many doors or windows of a specific type are located on a level. Repeat this exercise until you are comfortable with getting the correct answer.

9. Close without saving.

Exercise 8-2

Add Notes

AUTODESK CERTIFIED USER

Drawing Name: **notes.rvt**
Estimated Time to Completion: 10 Minutes

Scope
Use Annotation Symbols.

Solution

1. Activate the **Insert** ribbon.

   Select **Load Family** from the Load from Library panel.

   Load
   Family

   | File name: | "General Project Notes" "Floor Plan Notes - Embry Home" |
   | Files of type: | All Supported Files (*.rfa, *.adsk) |

   Hold down the Control key to select both files.

2. Locate the two annotation symbols created called *General Project Notes* and *Floor Plan Notes - Embry Home*.
   Press **Open**.

3.  Activate the **Ground Floor Plan** sheet.

   Sheet List
   Sheets (all)
   A.1 - Ground Floor Plan
   A.2 - Main Floor Plans
   A.3 - Sections & Details
   A.4 - Cover Sheet

4. Activate the **Annotate** ribbon.
   Select the **Symbol** tool on the Symbol panel.

   Symbol

5.

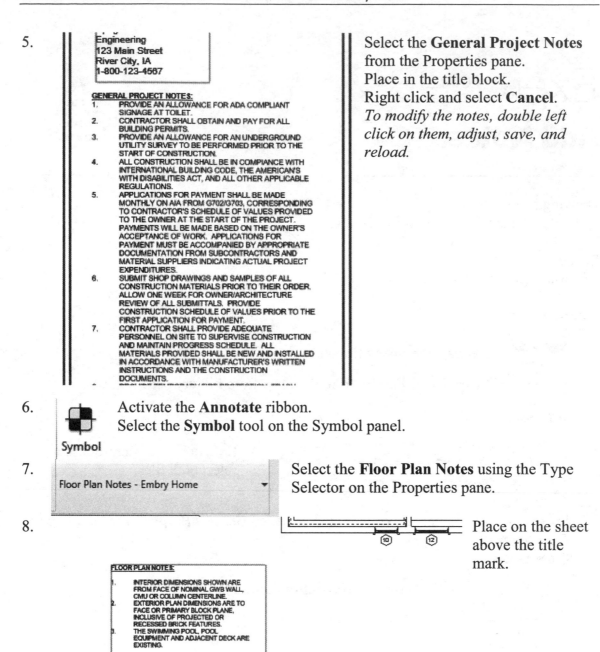

Engineering
123 Main Street
River City, IA
1-800-123-4567

GENERAL PROJECT NOTES:
1.   PROVIDE AN ALLOWANCE FOR ADA COMPLIANT
     SIGNAGE AT TOILET.
2.   CONTRACTOR SHALL OBTAIN AND PAY FOR ALL
     BUILDING PERMITS.
3.   PROVIDE AN ALLOWANCE FOR AN UNDERGROUND
     UTILITY SURVEY TO BE PERFORMED PRIOR TO THE
     START OF CONSTRUCTION.
4.   ALL CONSTRUCTION SHALL BE IN COMPIANCE WITH
     INTERNATIONAL BUILDING CODE, THE AMERICAN'S
     WITH DISABILITIES ACT, AND ALL OTHER APPLICABLE
     REGULATIONS.
5.   APPLICATIONS FOR PAYMENT SHALL BE MADE
     MONTHLY ON AIA FROM G702/G703, CORRESPONDING
     TO CONTRACTOR'S SCHEDULE OF VALUES PROVIDED
     TO THE OWNER AT THE START OF THE PROJECT.
     PAYMENTS WILL BE MADE BASED ON THE OWNER'S
     ACCEPTANCE OF WORK. APPLICATIONS FOR
     PAYMENT MUST BE ACCOMPANIED BY APPROPRIATE
     DOCUMENTATION FROM SUBCONTRACTORS AND
     MATERIAL SUPPLIERS INDICATING ACTUAL PROJECT
     EXPENDITURES.
6.   SUBMIT SHOP DRAWINGS AND SAMPLES OF ALL
     CONSTRUCTION MATERIALS PRIOR TO THEIR ORDER.
     ALLOW ONE WEEK FOR OWNER/ARCHITECTURE
     REVIEW OF ALL SUBMITTALS. PROVIDE
     CONSTRUCTION SCHEDULE OF VALUES PRIOR TO THE
     FIRST APPLICATION FOR PAYMENT.
7.   CONTRACTOR SHALL PROVIDE ADEQUATE
     PERSONNEL ON SITE TO SUPERVISE CONSTRUCTION
     AND MAINTAIN PROGRESS SCHEDULE. ALL
     MATERIALS PROVIDED SHALL BE NEW AND INSTALLED
     IN ACCORDANCE WITH MANUFACTURER'S WRITTEN
     INSTRUCTIONS AND THE CONSTRUCTION
     DOCUMENTS.

Select the **General Project Notes** from the Properties pane.
Place in the title block.
Right click and select **Cancel**.
To modify the notes, double left click on them, adjust, save, and reload.

6.   Activate the **Annotate** ribbon.
     Select the **Symbol** tool on the Symbol panel.

     Symbol

7.   Floor Plan Notes - Embry Home

     Select the **Floor Plan Notes** using the Type Selector on the Properties pane.

8.   Place on the sheet above the title mark.

FLOOR PLAN NOTES:
1.   INTERIOR DIMENSIONS SHOWN ARE
     FROM FACE OF NOMINAL GWB WALL,
     CMU OR COLUMN CENTERLINE.
2.   EXTERIOR PLAN DIMENSIONS ARE TO
     FACE OR PRIMARY BLOCK PLANE,
     INCLUSIVE OF PROJECTED OR
     RECESSED BRICK FEATURES.
3.   THE SWIMMING POOL, POOL
     EQUIPMENT AND ADJACENT DECK ARE
     EXISTING.

Administration Building Ground Floor
1/4" = 1'-0"

9.   Close without saving.

Exercise 8-3

Create a Schedule with Images

Drawing Name: **schedule_images.rvt**
Estimated Time to Completion: 10 Minutes

Scope
Use Images in Schedules.

You can produce schedules that include graphical information by associating images with elements in the building model. You can use imported images or saved views, such as renderings. The schedule view will display the image name, but not the image. The images will display when the schedule view is placed on a sheet.

Solution

1. Open the **Window Schedule**.

2. Select the **Edit** button next to Fields.

3. Locate the Image field and add it to the schedule.

Press **OK**.

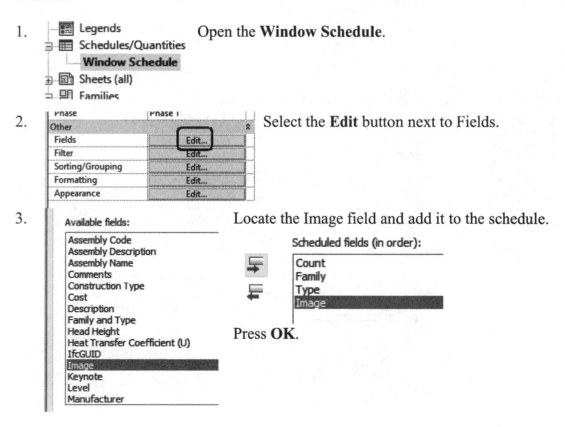

4.

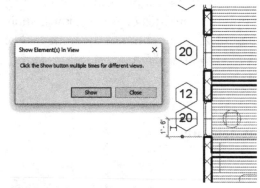

The schedule updates, but the column is empty.

5.

A	B	C
Count	Family	Type
5	Archtop with Trim	36" x 48"
7	Casement with Trim	16" x 24"
8	Casement with Trim	36" x 48"
12	Casement with Trim	36" x 72"
2	Fixed	36"w x 48"h

In the schedule, highlight the **Archtop with Trim** window.
Select **Highlight in Model** from the ribbon.

6.

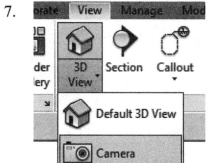

Click on the **Show** button until you get to a plan view.
Then close the dialog box.

7.

Go to the **View** ribbon.
Select the **Camera** tool under 3D View.

8.

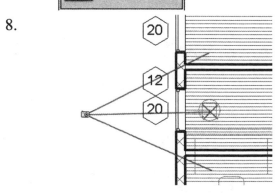

Position the camera to look directly at the window.

9. 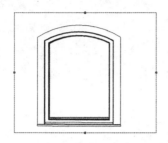 The camera should be at the 17' 6" elevation.

10.  Adjust the Crop Region to display only one window.

11. Turn off the visibility of the Crop Region.

12. Set the View Display to **Hidden Line**.

13. Locate the view created in the browser.
Right click and select **Save to Project as Image**.

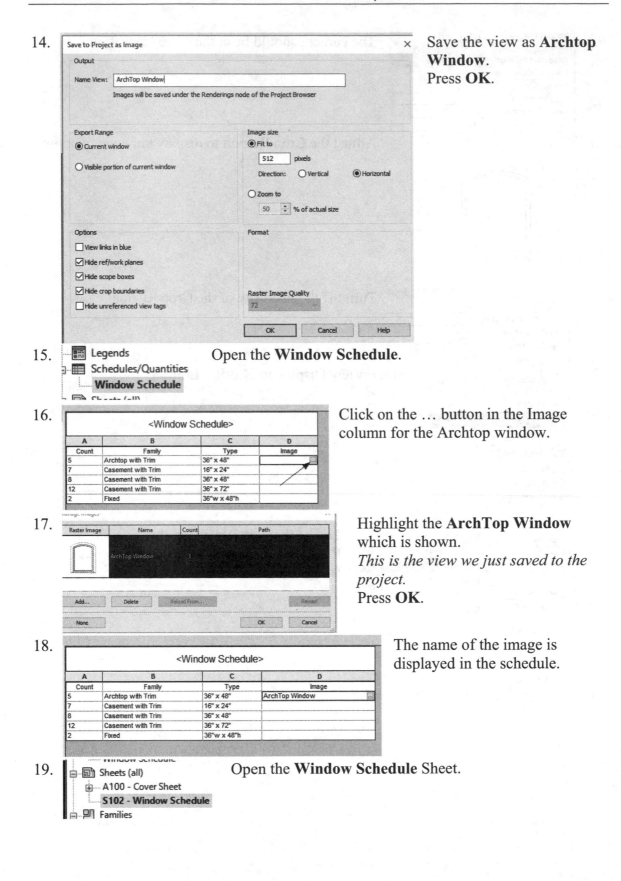

14. Save the view as **Archtop Window**.
Press **OK**.

15. Open the **Window Schedule**.

16. Click on the … button in the Image column for the Archtop window.

17. Highlight the **ArchTop Window** which is shown.
This is the view we just saved to the project.
Press **OK**.

18. The name of the image is displayed in the schedule.

19. Open the **Window Schedule** Sheet.

20.

Count	Family	Type	Image
5	Archtop with Trim	36" x 48"	
12	Casement Dbl with Trim	32" x 48"	
7	Casement with Trim	16" x 24"	
8	Double Hung with Trim	16" x 24"	
2	Fixed	36"w x 48"h	

Window Schedule

Drag and drop the window schedule onto the sheet.
The image is displayed.

21.
- Legends
- Schedules/Quantities
  - **Window Schedule**
- Sheets (all)

Open the **Window Schedule**.

22.

<Window Schedule>

	A	B	C	D
	Count	Family	Type	Image
5		Archtop with Trim	36" x 48"	Archtop Window
12		Casement Dbl with Trim	32" x 48"	
7		Casement with Trim	16" x 24"	
8		Double Hung with Trim	16" x 24"	
2		Fixed	36"w x 48"h	

Select the ... button in the Image column for the Casement with Trim window.

23.

Add...

Select the **Add** button.

24.

File name: Casement Window
Files of type: All Image Files (*.bmp, *.jpg, *.jpeg, *.png, *.tif)

Locate the **Casement Window** image file. Press **Open**.
This is included in the exercise files.

25.

Manage Images

Raster Image	Name	Count	Path
	ArchTop Window	6	
	Casement Window.JPG	0	D:\Revit Certification Guide 2017\Revit Advanced 2017 Exercises\Casement Window.JPG

The image is listed. Highlight and press **OK**.

26.

<Window Schedule>

	A	B	C	D
	Count	Family	Type	Image
5		Archtop with Trim	36" x 48"	Archtop Window
12		Casement Dbl with Trim	32" x 48"	
7		Casement with Trim	16" x 24"	Casement Window.JPG
8		Double Hung with Trim	16" x 24"	
2		Fixed	36"w x 48"h	

The schedule updates to show the image file name.

27.
- Window Schedule
- Sheets (all)
  - A100 - Cover Sheet
  - **S102 - Window Schedule**
- Families

Open the **Window Schedule** Sheet.

28.

Window Schedule			
Count	Family	Type	Image
5	Archtop with Trim	36" x 48"	
12	Casement Dbl with Trim	32" x 48"	
7	Casement with Trim	16" x 24"	
8	Double Hung with Trim	16" x 24"	
2	Fixed	36"w x 48"h	

The sheet has updated with the new image.

29. Legends
Schedules/Quantities
Window Schedule
Sheets (all)

Open the **Window Schedule**.

30.

		<Window Schedule>	
A	**B**	**C**	
Count	Family	Type	
5	Archtop with Trim	36" x 48"	ArchTop
7	Casement with Trim	16" x 24"	Caseme
8	Casement with Trim	36" x 48"	
12	Casement with Trim	36" x 72"	
2	Fixed	36"w x 48"h	

Highlight the Fixed Window in the schedule.

31. Highlight in Model

Element

Select **Highlight in Model**.

32.

When a plan view is displayed, close the Show dialog.

Show Element(s) In View

Click the Show button multiple times for different views.

Show   Close

33. Edit Family

Mode

Select the fixed window.
Select **Edit Family** from the ribbon.

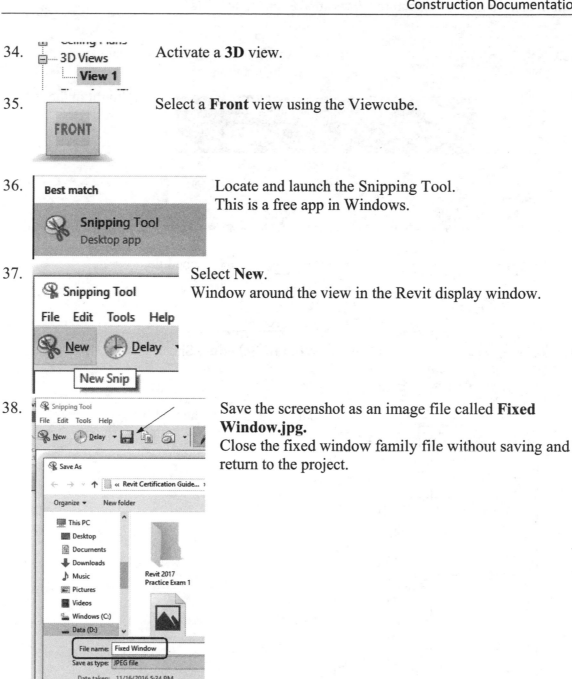

34. Activate a **3D** view.

35. Select a **Front** view using the Viewcube.

36. Locate and launch the Snipping Tool.
This is a free app in Windows.

37. Select **New**.
Window around the view in the Revit display window.

38. Save the screenshot as an image file called **Fixed Window.jpg.**
Close the fixed window family file without saving and return to the project.

39. Open the **Window Schedule**.

40. Select the ... button in the image column for the Fixed window.

41. Select the **Add** button.

42. **File name:** | Fixed Window

Locate the Fixed Window image file created with the snipping tool. Press **Open**.

43.

Fixed Window.JPG   0   D:\Revit Certification Guide 2017\Revit Advanced 2017 Exercises\Fixed Window.JPG

Highlight in the list and press **OK**.

| Add... | Delete | Reload From... | | Reload |

| None | | | OK | Cancel |

44.

<Window Schedule>

A	B	C	D
Count	Family	Type	Image
5	Archtop with Trim	36" x 48"	Archtop Window
12	Casement Dbl with Trim	32" x 48"	
7	Casement with Trim	16" x 24"	Casement Window.JPG
8	Double Hung with Trim	16" x 24"	
2	Fixed	36"w x 48"h	Fixed Window.JPG

The image file is listed in the schedule.

45.

- Sheets (all)
  - A100 - Cover Sheet
  - **S102 - Window Schedule**
- Families

Open the **Window Schedule** Sheet.

46.

Window Schedule

Count	Family	Type	Image
5	Archtop with Trim	36" x 48"	
12	Casement Dbl with Trim	32" x 48"	
7	Casement with Trim	16" x 24"	
8	Double Hung with Trim	16" x 24"	
2	Fixed	36"w x 48"h	

The sheet has updated with the new image.

47. Close without saving.

Extra: Add images to the remaining two rows in the schedule by creating and linking new image files.

Exercise 8-4

Creating a Legend

Drawing Name: **i_Legends.rvt**
Estimated Time to Completion: 30 Minutes

Scope
Create a Legend.
Add to a sheet.
Export Legend.

Solution

1. Activate the **View** tab on the ribbon.
    Select **Legend** from the Create panel.

2. Set the Name to **Door and Window Legend**.
    Set the Scale to **¼" = 1'-0"**.
    Press **OK**.

    Name: Door and Window Legend
    Scale: 1/4" = 1'-0"
    Scale value 1: 48

3. The Legend is listed in the browser. An empty view window is opened.

    Legends
    Door and Window Legend

4. In the browser, locate all the door families.

    Detail Items
    Doors
    Cased Opening
    double glass
    Sgl Flush
    single flush

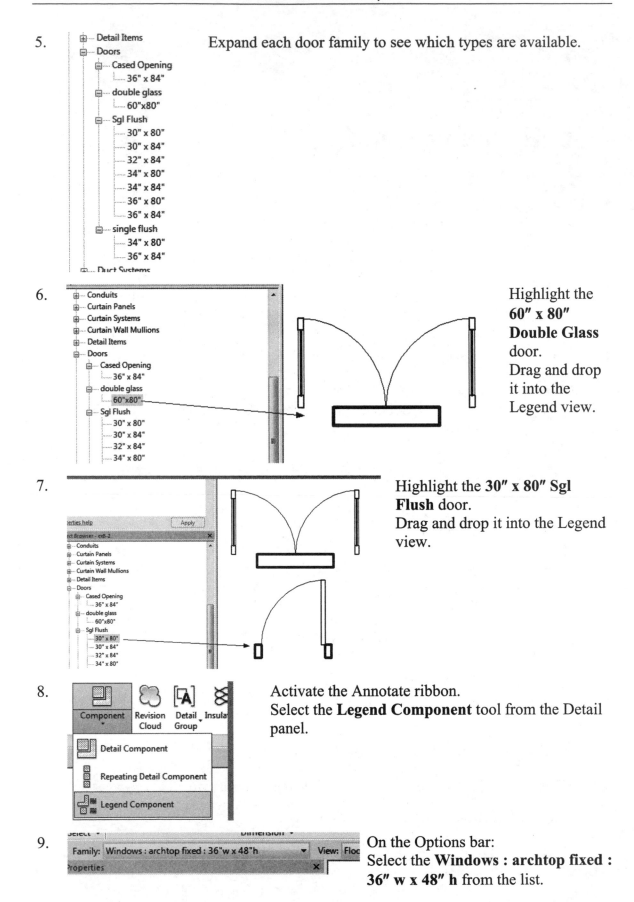

5. Expand each door family to see which types are available.

6. Highlight the **60″ x 80″ Double Glass** door.
   Drag and drop it into the Legend view.

7. Highlight the **30″ x 80″ Sgl Flush** door.
   Drag and drop it into the Legend view.

8. Activate the Annotate ribbon.
   Select the **Legend Component** tool from the Detail panel.

9. On the Options bar:
   Select the **Windows : archtop fixed : 36″ w x 48″ h** from the list.

10. 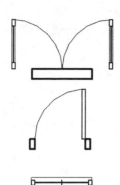 Place the symbol below the other two symbols.

11. 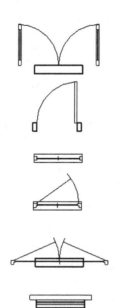 Drag and drop each window symbol into the Legend view.

Symbols can be dragged and dropped from the Project Browser or using the Legend Component tool.

12. **A**
Text

Activate the **Annotate** ribbon.
Select the **Text** tool.

13. DOUBLE GLASS DOOR   Add text next to each symbol.

SINGLE FLUSH DOOR

14. Legend Components : Legend Component - Windows :
archtop fixed : 36"w x 48"h (Floor Plan)

Mouse over a symbol to see what family type it is, if needed.

15.

DOUBLE GLASS DOOR

SINGLE FLUSH DOOR

ARCHTOP FIXED

CASEMENT

DOUBLE CASEMENT WITH TRIM

DOUBLE HUNG WITH TRIM

FIXED

Add text as shown.

16.

Detail
Line

Select the **Detail Line** tool from the Detail panel on the Annotate ribbon.

17.

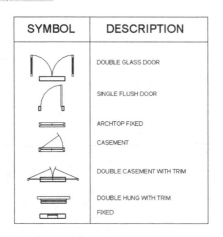

SYMBOL	DESCRIPTION
	DOUBLE GLASS DOOR
	SINGLE FLUSH DOOR
	ARCHTOP FIXED
	CASEMENT
	DOUBLE CASEMENT WITH TRIM
	DOUBLE HUNG WITH TRIM
	FIXED

Add a rectangle.
Add a vertical and horizontal line as shown.
Add the header text.

18.

Schedules/Quantities
Sheets (all)
A100
New Sheet...
Familie
Browser Organization...
Anno
Cable
Search...
Ceilings

Locate the **Sheets** category in the browser.
Right click and select **New Sheet**.

19.

Load...

Select **Load**.

20.

Titleblocks

Name
A 8.5 x 11 Vertical
B 11 x 17 Horizontal
C 17 x 22 Horizontal
D 22 x 34 Horizontal
E 34 x 44 Horizontal
E1 30 x 42 Horizontal

Locate the *Titleblocks* folder.
Locate the **D 22 x 34 Horizontal** title block.
Press **Open**.

21. Select titleblocks:

    D 22 x 34 Horizontal
    E1 30 x 42 Horizontal : E1 30x42 Horizontal
    None

    Highlight the **D 22 x 34 Horizontal** title block.
    Press **OK**.

22. Drag and drop the Level 1 floor plan on to the sheet.

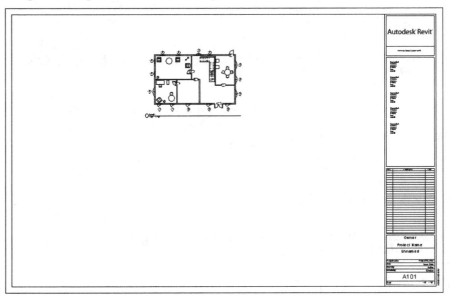

23. Graphics

View Scale	1/4" = 1'-0"
Scale Value 1:	48
Display Model	Normal

    In the Properties pane:
    Change the View Scale to ¼″ = 1′-0″.

24. Drag and drop the Door and Window Legend onto the sheet.
    Place next to the floor plan view.

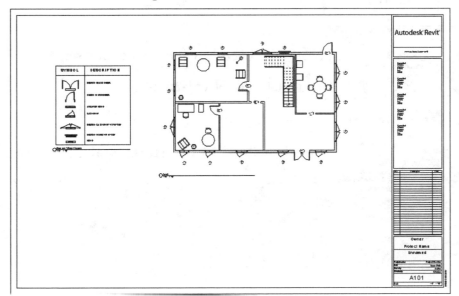

25. Select the legend so it highlights.

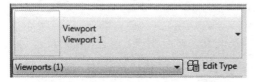

Select **Edit Type** on the Properties pane.

26. Duplicate... Select **Duplicate**.

27. Name: Viewport with no Title Type **Viewport with no Title**. Press **OK**.

28.

Type Parameters	
Parameter	
Graphics	
Title	view title
Show Title	No
Show Extension Line	☐
Line Weight	1
Color	Black
Line Pattern	Solid

Uncheck **Show Extension Line**.
Set Show Title to **No**.
Press **OK**.

29.

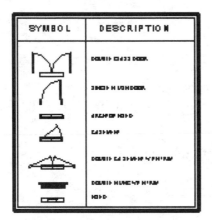

Zoom in to review the legend.

30.

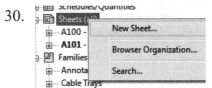

Locate the **Sheets** category in the browser.

Right click and select **New Sheet**.

31. Select titleblocks:

D 22 x 34 Horizontal
E1 30 x 42 Horizontal : E1 30x42 Horizontal
None

Highlight the **D 22 x 34 Horizontal** title block.

Press **OK**.

32. Drag and drop the **Level 2** floor plan on to the sheet.

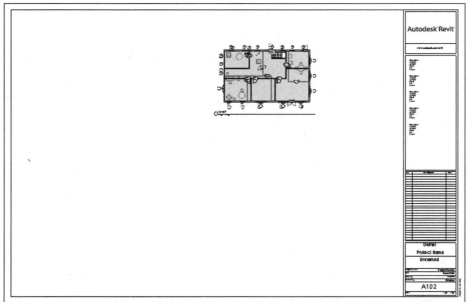

33.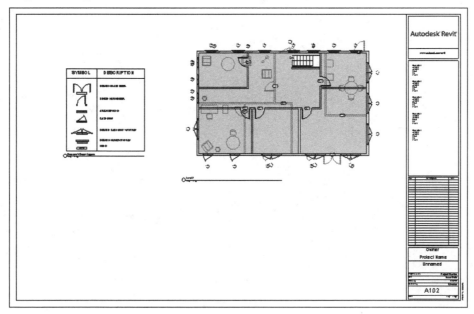

In the Properties pane:
Change the View Scale to **¼″ = 1′-0″**.

34. Drag and drop the Door and Window Legend onto the sheet. Place next to the view.

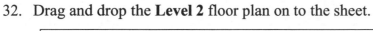

35. Select the legend view and then use the type selector in the Properties pane to set whether the view should have a title or not.

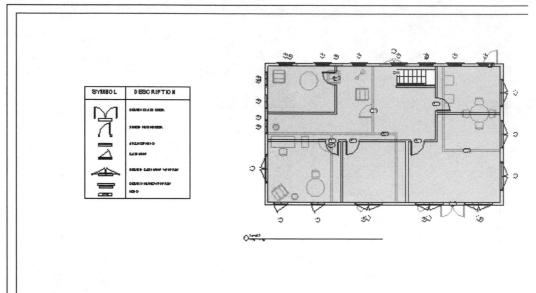

Note that legends can be placed on more than one sheet.

36. Save as *ex8-2.rvt*.

Exercise 8-5
Copying a Legend to a New Project

Drawing Name: **legend source.rvt, legend target.rvt**
Estimated Time to Completion: 10 Minutes

Scope
Copy a legend from one project to another

Solution

1.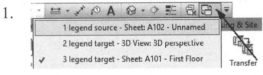

   Open the legend source and the legend target projects. Use the **Switch Windows** tool on the standard toolbar to switch to the *legend source* project.

2.

   Open the **Door and Window Legend** view in the *legend source* project.

3.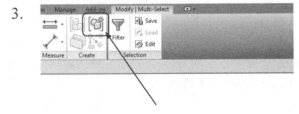

   Window around the entire legend to select. Select the **Create Group** tool on the ribbon.

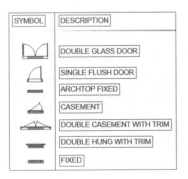

4.

Name: Door and Window Legend

☐ Open in Group Editor

[OK]   [Cancel]   [Help]

Name the group **Door and Window Legend**.
Press **OK**.

5.

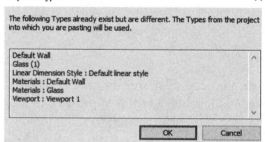

Scroll down the Project Browser.
Locate the group under Detail in the Groups category.
Right click and select **Copy to Clipboard**.

6.

Duplicate Types   ×

The following Types already exist but are different. The Types from the project into which you are pasting will be used.

Default Wall
Glass (1)
Linear Dimension Style : Default linear style
Materials : Default Wall
Materials : Glass
Viewport : Viewport 1

[OK]   [Cancel]

Use the **Switch Windows** tool to switch to the legend target project.
Press **Ctl+V** to paste the group into the *legend target* project.
A dialog advising of duplicate types may appear.
Press **OK**.

7.

⊞ 💷 Families
⊟ 🔲 Groups
  ⊟ Detail
     🔲 Door and Window Legend
  ⋯ Model
     Revit Links

The legend will appear under Groups in the legend target project.

8.

Autodesk Revit 2017 - Not For Resale Version -   ex8-3b - Sheet: A101 - First

View   Manage   Add-Ins   Modify

3D View   Section   Callout   Plan Views   Elevation   Drafting View   Duplicate View   Legends   S

Create   Legend

Go to the **View** ribbon.
Select the **Legend** tool.

9.

New Legend View   ×

Name: Door and Window Legend

Scale: 1/4" = 1'-0"

Scale value 1:   48

[OK]   [Cancel]

Type **Door and Window Legend**.
Press **OK**.

10.

SYMBOL	DESCRIPTION
	DOUBLE GLASS DOOR
	SINGLE FLUSH DOOR
	ARCHTOP FIXED
	CASEMENT
	DOUBLE CASEMENT WITH TRIM
	DOUBLE HUNG WITH TRIM
	FIXED

A blank view opens. Drag and drop the Door and Window Legend group into the view. Left click to place. Right click to Cancel out of the command.

11. Select the group that was placed.
Select **Ungroup** from the ribbon.

12. Change the View Scale of the Legend to **1/8" = 1' -0"**.

Legend	

Legend: Door and Window Legend	Edit Type
Graphics	
View Scale	1/8" = 1'-0"
Scale Value 1:	96
Detail Level	Coarse

13. Window around the legend. Select **Filter**. Select the **Text Notes**. Press **OK**.

Category:	Count:
☐ Legend Components	7
☐ Lines (Thin Lines)	10
☑ Text Notes	8

14. Text
3/32" text

Use the Type Selector to change the text notes to use **3/32" text**.

15. Schedules/Quantities
Sheets (all)
A101 - First Floor
Families

Activate the **A101 – First Floor** sheet under *Sheets*.

16.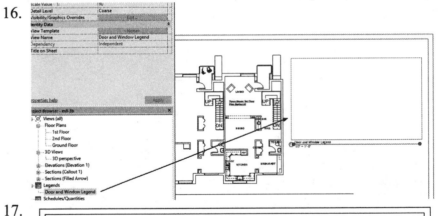

Drag and drop to place on the sheet.

17.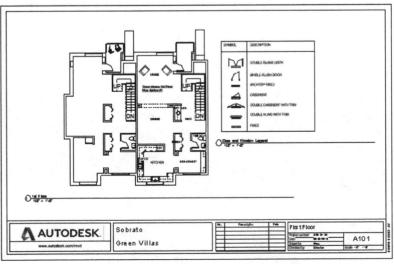

Close without saving.

Exercise 8-6
Adding Rooms to a Floor Plan

Drawing Name: **i_rooms.rvt**
Estimated Time to Completion: 10 Minutes

Scope
Add rooms to a floor plan.

Solution

1. Activate the **2ⁿᵈ Floor** floor plan.

   - Floor Plans
     - 1st Floor
     - **2nd Floor**
     - Ground Floor

2. Select the **Room** tool from the Room & Area panel on the Architecture ribbon.

   Room

3. Place a room in each enclosed boundary.

4. 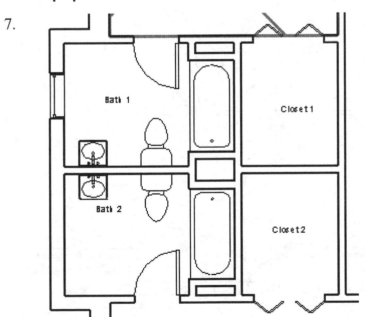 Select the room in the upper left corner.

5.

Identity Data	
Number	8
Name	Master Bedroom
Image	
Comments	
Occupancy	2
Department	

In the Properties pane:
Enter the data as shown.
Press **Apply**.

6.

Master Bedroom

Note that the Room Name updates to the name assigned.

7.

Bath 1

Closet 1

Bath 2

Closet 2

Add the room names shown.

8.   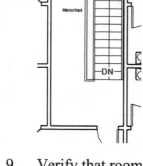 Continue adding names to the rooms.

9.   Verify that room names have been assigned to all rooms.

10.   Save as *ex8-3.rvt*.

Exercise 8-7

Creating an Area Scheme

Drawing Name: **i_Color_Scheme.rvt**
Estimated Time to Completion: 15 Minutes

Scope
Create an area plan using a color scheme.
Place a color scheme legend.

Solution

1. Activate the **01- Entry Level** floor plan.

   Floor Plans
   — **01 - Entry Level**
   — 01 - Entry Level-Rooms
   — 02 - Floor
   — 03 - Floor
   — Roof

2. In the Properties pane:
   Scroll down to the **Color Scheme** field.
   Press **<none>**.

Wall Join Display	Clean all wall joins
Discipline	Architectural
Color Scheme Lo...	Background
Color Scheme	<none>
Default Analysis ...	None

3. Select the **Rooms** category.
   Highlight **Name**.

   Schemes
   Category:
   Rooms

   (none)
   Name
   Department

4. In the Title field, enter **Area Legend**.
   Under Color: select the **Area** scheme.

   Scheme Definition
   Title: Area Legend
   Color: Area

5. Press **OK**.

   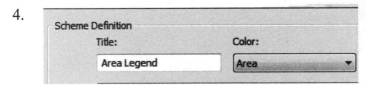

   Colors are not preserved when changing which parameter is colored. To color by a different parameter, consider making a new color scheme.

6.

	Value	Visible	Color	Fill Pattern	Preview
1	47 SF	☑	RGB 156-185	Solid fill	
2	58 SF	☑	PANTONE 3	Solid fill	
3	64 SF	☑	PANTONE 6	Solid fill	
4	76 SF	☑	RGB 139-166	Solid fill	
5	90 SF	☑	PANTONE 6	Solid fill	
6	94 SF	☑	RGB 096-175	Solid fill	
7	95 SF	☑	RGB 209-203	Solid fill	
8	133 SF	☑	RGB 173-118	Solid fill	
9	168 SF	☑	RGB 194-161	Solid fill	

Note that different colors are applied depending on the square footage of the rooms.

Press **OK**.

7.

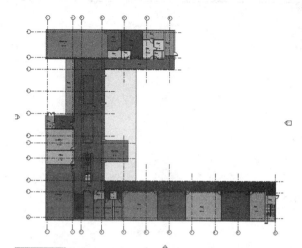

The rooms fill in according to the square footage.

8.

Activate the **Annotate** ribbon.
Select the **Color Fill Legend** tool on the Color Fill panel.

9.

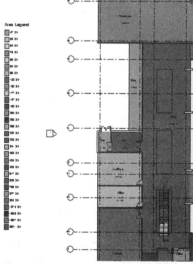

Place the legend in the view.

10.

Select the legend that was placed in the display window.
Select **Edit Scheme** from the Scheme panel on the ribbon.

11. Change the Fill Pattern for 47 SF **to Crosshatch- Small**.

	Value	Visible	Color	Fill Pattern	Preview	
1	47 SF	☑	RGB 156-185	Crosshatch-small ▼		
2	58 SF	☑	PANTONE 3	Solid fill		
3	64 SF	☑	PANTONE 6	Solid fill		

12. Change the hatch patterns for the next few rows. Press **OK**.

	Value	Visible	Color	Fill Pattern	Preview	
1	47 SF	☑	RGB 156-185	Crosshatch-small		
2	58 SF	☑	PANTONE 3	Crosshatch		
3	64 SF	☑	PANTONE 6	Diagonal crosshatch		
4	76 SF	☑	RGB 139-166	Diagonal down ▼		
5	90 SF	☑	PANTONE 6	Solid fill		
6	94 SF	☑	RGB 096-175	Solid fill		

13. Area Legend   *Note that the legend updates.*

   47 SF
   58 SF
   64 SF
   76 SF
   90 SF

14. Color Fill Legend
    1

    gends (1)   Edit Type

   Select the Color Fill Legend. On the Properties pane:
   Select **Edit Type**.

15. **Text**

Font	Tahoma
Size	3/16"
Bold	☑
Italic	☐
Underline	☐

   Change the font for the Text to **Tahoma.**
   Enable **Bold**.

16. **Title Text**

Font	Tahoma
Size	1/4"
Bold	☑
Italic	☑
Underline	☐

   Change the font for the Title Text to **Tahoma.**
   Enable **Bold**.
   Enable **Italic.**
   Press **OK**.

17. *Area Legend*

   47 SF
   58 SF
   64 SF

   Note how the legend updates.

18. Close without saving.

Exercise 8-8
Creating an Area Plan

Drawing Name: **i_Area_Plan.rvt**
Estimated Time to Completion: 30 Minutes

Scope
Create an area plan.
Add areas to a floor plan.

Solution

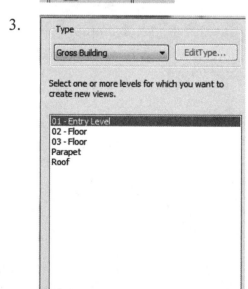

1. 　Floor Plans
　　　01 - Entry Level
　　　02 - Floor
　　　03 - Floor
　　　Roof
　　　Site

　Activate the **01- Entry Level** floor plan.

2. Area | Area Boundary | Tag Area
　Area Plan

　Select the **Area Plan** tool from the Room & Area panel on the **Architecture** ribbon.

3. Type
　Gross Building ▼ EditType...

　Select one or more levels for which you want to create new views.

　01 - Entry Level
　02 - Floor
　03 - Floor
　Parapet
　Roof

　☑ Do not duplicate existing views

　Select **Gross Building** under the Type drop-down.
　Select **01- Entry Level**.
　Enable **Do not duplicate existing views**.
　Press **OK**.

4. Automatically create area boundary lines associated with external walls and gross building area?

Press **Yes**.

5. ```
   └── Section 2
   └── Area Plans (Gross Building)
       └── 01 - Entry Level
   Legends
   ```

A new area plan will be listed in the Project Browser.

6.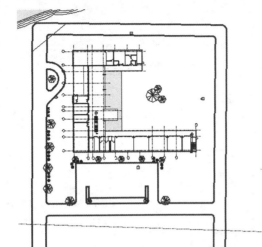

A new view will open in the graphics window.

7. Type **VV** to launch the Visibility/Graphics dialog.

8. ```
   ⊞ ☑ Structural Framing
   ⊞ ☐ Topography
   ⊞ ☑ Walls
   ⊞ ☑ Windows
   ```

Clear the **Topography** box to turn off the visibility.

9. ```
   ⊞ ☑ Mechanical Equipm...
   ⊞ ☑ Parking
   ⊞ ☐ Parts
   ⊞ ☐ Planting          Overr
   ⊞ ☑ Plumbing Fixtures
   ⊞ ☑ Railings
   ⊞ ☑ Ramps
   ```

Clear the **Planting** box to turn off the visibility.
Press **OK** to close the dialog.

10.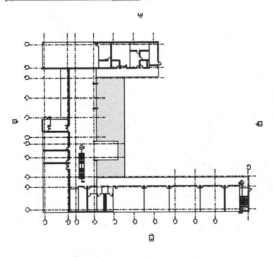

The view now shows just the building with grid lines.
   ```

11.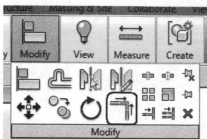

Activate the **Modify** ribbon.
Select the **Trim** tool.

12.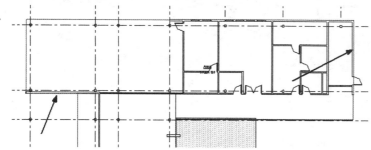

Select the two lines indicated.

13.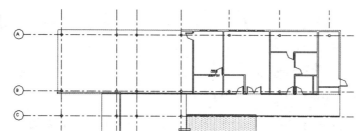

The boundary lines trim to create an enclosed area.

14.

Area Boundary

Activate the **Architecture** ribbon.
Select the **Area Boundary** tool.

15.

Draw

Select the **Pick Lines** tool from the Draw panel.

16. Select the wall indicated to place an area boundary line.
This divides the area into two different sections.

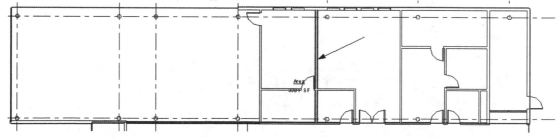

17. Activate the **Architecture** ribbon.
Select the **Area** tool.
Place an area in the area on the right.

18. You should see two tags. One is for the area on the left and one is for the area on the right. You may need to move them to see them clearly.

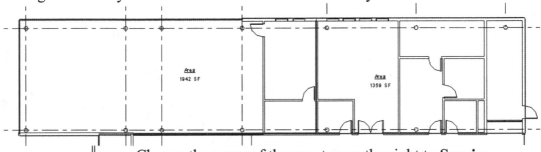

Area
1942 SF

Area
1359 SF

19. Change the name of the area tag on the right to **Service**.
Left click on the text to edit.

Service
1359 SF

20. Change the label for the left area tag to **Administration**.

dministration
1942 SF

21.

Wall Join Display	Clean all wall joins
Discipline	Architectural
Color Scheme Lo...	Background
Color Scheme	<none>
Default Analysis ...	None
Sun Path	

Left click anywhere in the window to clear any selections.
On the Properties pane:
Click on **<none>** next to Color Scheme.

22.

Schemes
Category:
Areas (Gross Building) ▼

(none)
Gross Building Area

Select **Gross Building Area** under Category.

23.

Scheme Definition
Title: Color:
Building Area Legend   Name ▼

Under Color: select **Name**.

24.

Colors are not preserved when changing which parameter is colored. To color by a different parameter, consider making a new color scheme.
OK   Cancel

Press **OK**.

25.

	Value	Visible	Color	Fill Pattern
1	Administration	☑	RGB 156-185	Solid fill
2	Service	☑	PANTONE 3	Solid fill

The two named areas are listed.

Colors are automatically assigned.

26. Press **OK**.
27. Save as *ex8-5.rvt*.

Exercise 8-9
Creating a Room Finish Schedule

Drawing Name: **i_Finish_Schedule.rvt**
Estimated Time to Completion: 10 Minutes

Scope
Create a room finish schedule.

Solution

1.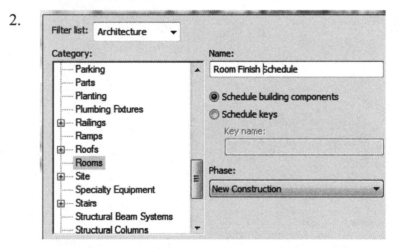
   Activate the **View** tab on the ribbon.
   Select the **Schedule/Quantities** tool from the Create panel.

2. Select **Rooms** in the Category list box.
   Change the Name to **Room Finish Schedule**.
   Press **OK**.

3. Add the fields shown to the schedule.
   Use the Control key to select more than one field.
   Move them over using the arrow key.
   Use the Up and Down arrows to organize.

   Scheduled fields (in order):
   Number
   Name
   Base Finish
   Floor Finish
   Wall Finish
   Ceiling Finish
   Area
   Comments
   Level

4.

Select the **Sorting/ Grouping** tab. Select **Number** from the drop-down.

5.

Select the **Formatting** tab. Highlight the Base Finish field. Change the Heading to **Base**.

6.

Highlight the Floor Finish field. Change the Heading to **Floor**.

7.

Highlight the Wall Finish field. Change the Heading to **Wall**.

8.

Highlight the Ceiling Finish field. Change the Heading to **Ceiling**.

9. Press **OK** to finish the schedule.

10. The schedule view opens.

<Room Finish Schedule>

A	B	C	D	E	F	G	H	I
Number	Name	Base	Floor	Wall	Ceiling	Area	Comments	Level
101	Vest.					406 SF		01 - Entry Level
102	Lobby					3537 SF		01 - Entry Level
103	Conference					359 SF		01 - Entry Level
104	Instruction					712 SF		01 - Entry Level
105	Instruction					891 SF		01 - Entry Level
106	Instruction					712 SF		01 - Entry Level
107	Corridor					1504 SF		01 - Entry Level
108	Instruction					900 SF		01 - Entry Level
109	Women					135 SF		01 - Entry Level

11. Drag your cursor to select the Base, Floor, Wall, and Ceiling headers. Select the **Group** tool.

Group

12. Add the word **FINISH** above the grouped fields.

<Room Finish Schedule>

B	C	D	E	F	G
		FINISH			
Name	Base	Floor	Wall	Ceiling	Area
Vest.					406 SF
Lobby					3537 SF
Conference					359 SF
Instruction					712 SF

13. To add finish information, just type in the cell.

	B	C	D	E	F	G
			FINISH			
er	Name	Base	Floor	Wall	Ceiling	Area
	Vest.	PRIMER	GRANITE	SW 6136	SW 6136	406 SF
	Lobby					3537 SF
	Conference					359 SF
	Instruction					712 SF
	Instruction					891 SF
	Instruction					712 SF

14. Close the file without saving.

Exercise 8-10
Creating a Drawing List

Drawing Name: **sheets.rvt**
Estimated Time to Completion: 15 Minutes

Scope
Create a sheet list schedule.
Add to a sheet.

Solution

1.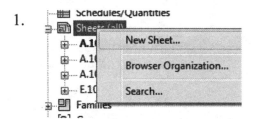

    Highlight **Sheets** in the Project Browser.
    Right click and select **New Sheet**.

2. Load...

    Select **Load**.

3. Titleblocks

    Name

    A 8.5 x 11 Vertical
    B 11 x 17 Horizontal
    C 17 x 22 Horizontal
    D 22 x 34 Horizontal
    E 34 x 44 Horizontal
    E1 30 x 42 Horizontal

    Browse to the *Titleblocks* folder.
    Select the *B 11 x 17 Horizontal Titleblock* and press **Open**.

4. Select titleblocks:

    B 11 x 17 Horizontal
    E1 30 x 42 Horizontal : E1 30x42 Horizontal
    None

    The B size Titleblock is listed in the dialog list.
    Highlight and press **OK**.

5. | | |
    |---|---|
    | Checked By | Checker |
    | Sheet Number | 0.0 |
    | Sheet Name | Cover Sheet |
    | Sheet Issue Date | 09/23/13 |

    In the Properties pane:
    Type **0.0** for the Number.
    Type **Cover Sheet** for the Sheet Name.

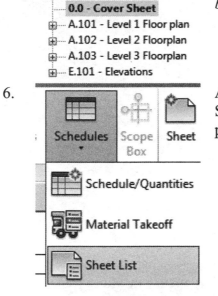

····· Schedules/Quantities
⊟··🔂 Sheets (all)
    ··· **0.0 - Cover Sheet**
    ⊞··· A.101 - Level 1 Floor plan
    ⊞··· A.102 - Level 2 Floorplan
    ⊞··· A.103 - Level 3 Floorplan
    ⊞··· E.101 - Elevations

Notice that sheet name and number update in the browser.

6. Activate the View ribbon.
Select the **Sheet List** tool under Schedules on the Create panel.

7. Add the following fields to the schedule:

Scheduled fields (in order):

Sheet Number
Sheet Name
Current Revision
Sheet Issue Date

Sheet Number
Sheet Name
Current Revision
Sheet Issue Date

8. Select the Sorting/Grouping tab.
Sort by **Sheet Number**.

| Fields | Filter | Sorting/Grouping | Formatting | Appearance |

Sort by: Sheet Number ⦿ Ascending
☐ Header ☐ Footer:

9. Press **OK** to create the schedule.

10. The drawing list schedule view is displayed.

Sheet List			
Sheet Numbe	Sheet Name	Current Revis	Sheet Issue D
0.0	Cover Sheet		07/18/12
A.101	Level 1 Floor plan		07/18/12
A.102	Level 2 Floorplan		07/18/12
A.103	Level 3 Floorplan		07/18/12
E.101	Elevations		07/18/12

11. Activate the **0.0- Cover Sheet**.

····· Legends
⊞··🧾 Schedules/Quantities
⊟··🔂 Sheets (all)
    ····· **0.0 - Cover Sheet**
    ⊞··· A.101 - Level 1 Floor plan
    ⊞··· A.102 - Level 2 Floorplan
    ⊞··· A.103 - Level 3 Floorplan

12.

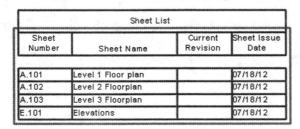

Drag and drop the sheet list onto the cover sheet.

13.

····· Sheet List
⊟····· Sheets (all)
   ⊞···· **0.0 - Cover Sheet**
   ⊞···· A.101 - Level 1 Floor plan
   ⊞···· A.102 - Level 2 Floorplan
   ⊞···· A.103 - Level 3 Floorplan
   ⊞···· E.101 - Elevations

Activate the Cover Sheet.

14.

Sheet Name	Cover Sheet
Sheet Issue Date	09/23/13
Appears In Sheet List	☐
Revisions on Sheet	Edit...
Other	⌃
File Path	
Drawn By	Author
Guide Grid	<None>

In the Properties pane:
Uncheck **Appears in Sheet List**.

Sheet List			
Sheet Number	Sheet Name	Current Revision	Sheet Issue Date
A.101	Level 1 Floor plan		07/18/12
A.102	Level 2 Floorplan		07/18/12
A.103	Level 3 Floorplan		07/18/12
E.101	Elevations		07/18/12

15. The schedule will automatically update.
Close without saving.

Exercise 8-11
Create a Note Symbol

Drawing Name: **none**
Estimated Time to Completion: 30 Minutes

Scope
Create Note Symbol.

Solution

1. 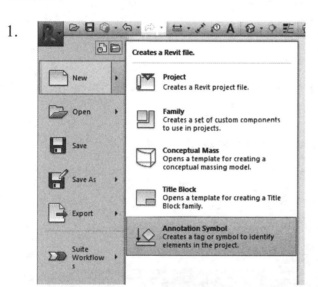   Go to **New→Annotation Symbol** on the Applications Menu.

2.

File name:	Generic Annotation
Files of type:	Family Template Files (*.rft)

Select *Generic Annotation*.
Press **Open**.

3. Locate the Word document titled *Notes* in the exercise folder and open it.

4. FLOOR PLAN NOTES:

   INTERIOR DIMENSIONS SHOWN ARE FROM FACE OF NOMINAL GWB WALL, CMU OR COLUMN CENTERLINE.

   EXTERIOR PLAN DIMENSIONS ARE TO FACE OR PRIMARY BLOCK PLANE, INCLUSIVE OF PROJECTED OR RECESSED BRICK FEATURES.

   THE SWIMMING POOL, POOL EQUIPMENT AND ADJACENT DECK ARE EXISTING.

   Highlight the Floor Plan Notes.
   Use Ctrl+C to copy.

5. Switch to Revit.

6.  Select the **Text** tool from the Text panel.
Place your cursor inside the text box.
Enter Ctrl+V to paste.

7. 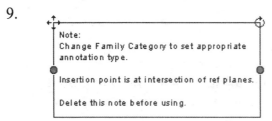 Use the Format settings to add an underline and bold the heading for the notes.

8. Left click anywhere in the drawing window to exit the Text command.
Right click and select **Cancel**.

9.  Select the note that was included in the template.
Right click and select **Delete**.

10. 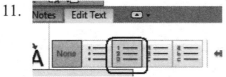 Position the note so it aligns with the intersection of the reference planes.

This is the insertion point.

11. 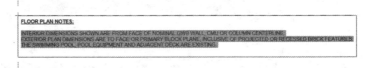 Highlight the text below the header.

Select the **Number** format.

12. 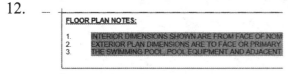 The numbers auto-insert and the text indents.

13. 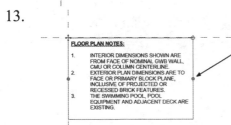 Left click outside the text box.
Use the right grip to re-size the text box.

14.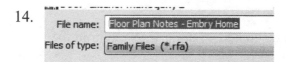

Save the family as *Floor Plan Notes - Embry Home*.

15. Close the file.

16.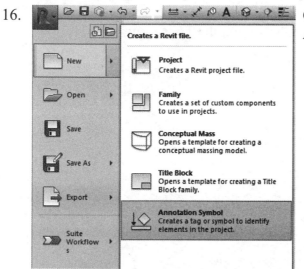

Go to **New→Annotation Symbol** on the Applications Menu.

17.

Select *Generic Annotation*.
Press **Open**.

18. Locate the Word document titled *Notes* in the exercise folder and open it.

19.

Highlight the General Project Notes.
Use Ctrl+C to copy.

20.

Switch to Revit.
Select the **Text** tool from the Text panel.

21.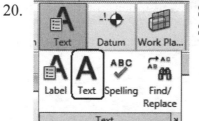

Place your cursor inside the text box.
Enter Ctrl+V to paste.

22. Use the Format settings to add an underline and bold the heading for the notes.

GENERAL PROJECT NOTES:
PROVIDE AN ALLOWANCE FOR A

23. Left click anywhere in the drawing window to exit the Text command. Right click and select **Cancel**.

24. Use the grips on the sides to resize the text box.

GENERAL PROJECT NOTES:

1. PROVIDE AN ALLOWANCE FOR ADA COMPLIANT SIGNAGE AT TOILET.
2. CONTRACTOR SHALL OBTAIN AND PAY FOR ALL BUILDING PERMITS.
3. PROVIDE AN ALLOWANCE FOR AN UNDERGROUND UTILITY SURVEY TO BE PERFORMED PRIOR TO THE START OF CONSTRUCTION.
4. ALL CONSTRUCTION SHALL BE IN COMPLIANCE WITH INTERNATIONAL BUILDING CODE, THE AMERICAN'S WITH DISABILITIES ACT, AND ALL OTHER APPLICABLE REGULATIONS.
5. APPLICATIONS FOR PAYMENT SHALL BE MADE MONTHLY ON AIA FROM G702/G703, CORRESPONDING TO CONTRACTOR'S SCHEDULE OF VALUES PROVIDED TO THE OWNER AT THE START OF THE PROJECT. PAYMENTS WILL BE MADE BASED ON THE OWNER'S ACCEPTANCE OF WORK. APPLICATIONS FOR PAYMENT MUST BE ACCOMPANIED BY APPROPRIATE DOCUMENTATION FROM SUBCONTRACTORS AND MATERIAL SUPPLIERS INDICATING ACTUAL PROJECT EXPENDITURES.
6. SUBMIT SHOP DRAWINGS AND SAMPLES OF ALL CONSTRUCTION MATERIALS PRIOR TO THEIR ORDER. ALLOW ONE WEEK FOR OWNER/ARCHITECTURE REVIEW OF ALL SUBMITTALS. PROVIDE CONSTRUCTION SCHEDULE OF VALUES PRIOR TO THE FIRST APPLICATION FOR PAYMENT.
7. CONTRACTOR SHALL PROVIDE ADEQUATE PERSONNEL ON SITE TO SUPERVISE CONSTRUCTION AND MAINTAIN PROGRESS SCHEDULE. ALL MATERIALS PROVIDED SHALL BE NEW AND INSTALLED IN ACCORDANCE WITH MANUFACTURER'S WRITTEN INSTRUCTIONS AND THE CONSTRUCTION DOCUMENTS.
8. PROVIDE TEMPORARY FIRE PROTECTION, TRASH REMOVAL, SECURITY, ELECTRICAL, AND WATER SERVICES. PROVIDE PORTABLE TEMPORARY TOILET FACILITIES.
9. CONTRACTORS MAY PROPOSE APPROVED EQUIVALENT PRODUCTS BY OTHER MANUFACTURERS FOR SUBSTITUTION IN LIEU OF SPECIFIED HEREIN.
10. PRIOR TO SUBSTANTIAL COMPLETION, THOROUGHLY CLEAN ALL BUILDING ELEMENTS AND SURFACES WITHIN THE AREA OF CONSTRUCTION. FOLLOWING A SUBSTANTIAL COMPLETION, PROVIDE A WRITTEN PUNCHLIST OF ITEMS TO THE OWNER TO BE COMPLETED WITHIN 15 DAYS.
11. PROVIDE SUBSTRATE SOIL TREATMENT EQUIVALENT TO AN EPA-REGISTERED TERMITICIDE. PROVIDE A PROPOSAL FOR CONTINUING SERVICE INCLUDING MONITORING INSPECTION AND RETREATMENT.

25. Use the **Numbers** formatting tool on the Format panel to add numbers and indent the paragraphs.

GENERAL PROJECT NOTES:

1. PROVIDE AN ALLOWANCE FOR ADA
2. CONTRACTOR SHALL OBTAIN AND I
3. PROVIDE AN ALLOWANCE FOR AN U
   TO BE PERFORMED PRIOR TO THE S
4. ALL CONSTRUCTION SHALL BE IN C
   BUILDING CODE, THE AMERICAN'S I
   OTHER APPLICABLE REGULATIONS.
5. APPLICATIONS FOR PAYMENT SHAL
   G702/G703, CORRESPONDING TO CC
   VALUES PROVIDED TO THE OWNER
   PAYMENTS WILL BE MADE BASED O
   WORK. APPLICATIONS FOR PAYMEI

26. Select the template note. Right click and select **Delete**.

Note:
Change Family Category to set appropriate annotation type.

Insertion point is at intersection of ref planes.

Delete this note before using.

27.

GENERAL PROJECT NOTES:

1. PROVIDE AN ALLOWANCE FOR ADA COMPLIANT SIGNAGE AT TOILET.
2. CONTRACTOR SHALL OBTAIN AND PAY FOR ALL BUILDING PERMITS.
3. PROVIDE AN ALLOWANCE FOR AN UNDERGROUND UTILITY SURVEY TO BE PERFORMED PRIOR TO THE START OF CONSTRUCTION.
4. ALL CONSTRUCTION SHALL BE IN COMPIANCE WITH INTERNATIONAL BUILDING CODE, THE AMERICAN'S WITH DISABILITIES ACT, AND ALL OTHER APPLICABLE REGULATIONS.
5. APPLICATIONS FOR PAYMENT SHALL BE MADE MONTHLY ON AIA FROM G702/G703, CORRESPONDING TO CONTRACTOR'S SCHEDULE OF VALUES PROVIDED TO THE OWNER AT THE START OF THE PROJECT. PAYMENTS WILL BE MADE BASED ON THE OWNER'S ACCEPTANCE OF WORK. APPLICATIONS FOR PAYMENT MUST BE ACCOMPANIED BY APPROPRIATE DOCUMENTATION FROM SUBCONTRACTORS AND MATERIAL SUPPLIERS INDICATING ACTUAL PROJECT EXPENDITURES.
6. SUBMIT SHOP DRAWINGS AND SAMPLES OF ALL CONSTRUCTION MATERIALS PRIOR TO THEIR ORDER. ALLOW ONE WEEK FOR OWNER/ARCHITECTURE REVIEW OF ALL SUBMITTALS. PROVIDE CONSTRUCTION SCHEDULE OF VALUES PRIOR TO THE FIRST APPLICATION FOR PAYMENT.
7. CONTRACTOR SHALL PROVIDE ADEQUATE PERSONNEL ON SITE TO SUPERVISE CONSTRUCTION AND MAINTAIN PROGRESS SCHEDULE. ALL MATERIALS PROVIDED SHALL BE NEW AND INSTALLED IN ACCORDANCE WITH MANUFACTURER'S WRITTEN INSTRUCTIONS AND THE CONSTRUCTION DOCUMENTS.
8. PROVIDE TEMPORARY FIRE PROTECTION, TRASH REMOVAL, SECURITY, ELECTRICAL, AND WATER SERVICES. PROVIDE PORTABLE TEMPORARY TOILET FACILITIES.
9. CONTRACTORS MAY PROPOSE APPROVED EQUIVALENT PRODUCTS BY OTHER MANUFACTURERS FOR SUBSTITUTION IN LIEU OF SPECIFIED HEREIN.
10. PRIOR TO SUBSTANTIAL COMPLETION, THOROUGHLY CLEAN ALL BUILDING ELEMENTS AND SURFACES WITHIN THE AREA OF CONSTRUCTION. FOLLOWING A SUBSTANTIAL COMPLETION, PROVIDE A WRITTEN PUNCHLIST OF ITEMS TO THE OWNER TO BE COMPLETED WITHIN 15 DAYS.
11. PROVIDE SUBSTRATE SOIL TREATMENT EQUIVALENT TO AN EPA-REGISTERED TERMITICIDE. PROVIDE A PROPOSAL FOR CONTINUING SERVICE INCLUDING MONITORING INSPECTION AND RETREATMENT.

The notes should look like this.

The width of the note should be no more than 5″ in order to fit properly in the title block.

28.

File name: General Project Notes

Files of type: Family Files (*.rfa)

Save the family as *General Project Notes*.

29. Close the file.

Exercise 8-12
Create a Material TakeOff Schedule

Drawing Name: **i_materials.rvt**
Estimated Time to Completion: 10 Minutes

Scope
Create a material takeoff schedule

Solution

1.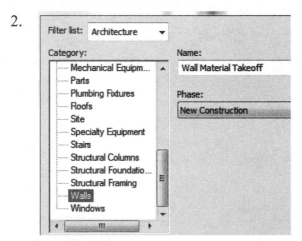
   Activate the View ribbon.
   Select **Schedules→Material Takeoff** from the Create panel.

2. Highlight **Walls**.
   Press **OK**.

3. On the **Fields** tab:

   Add the following fields:
   - Family and Type
   - Material: Name
   - Material: Area

4.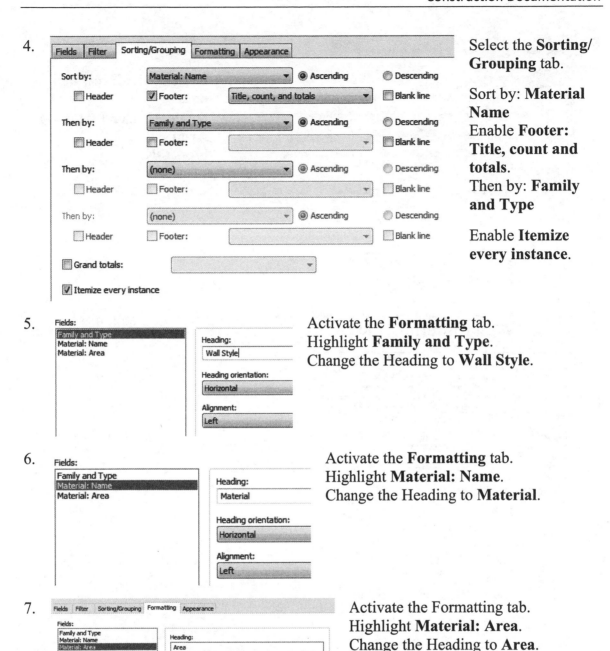

Select the **Sorting/Grouping** tab.

Sort by: **Material Name**

Enable **Footer: Title, count and totals**.

Then by: **Family and Type**

Enable **Itemize every instance**.

5.

Activate the **Formatting** tab.
Highlight **Family and Type**.
Change the Heading to **Wall Style**.

6.

Activate the **Formatting** tab.
Highlight **Material: Name**.
Change the Heading to **Material**.

7.

Activate the Formatting tab.
Highlight **Material: Area**.
Change the Heading to **Area**.
Select **Calculate totals** from the drop-down list.

8.

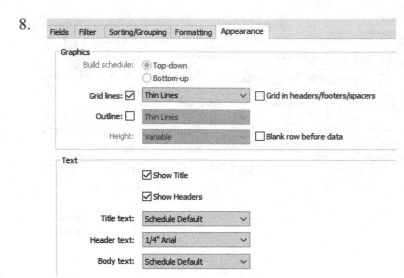

Activate the **Appearance** tab.
Disable **Blank row before data**.
Set the Header Text to **1/4″ Arial**.
Press **OK**.

9.

<Wall Material Takeoff>

A	B	C
Wall Style	Material	Area
Basic Wall: Exterior - Brick on Mtl. Stud	Air Barrier - Air Infiltration Barrier	914 SF
Basic Wall: Exterior - Brick on Mtl. Stud	Air Barrier - Air Infiltration Barrier	223 SF
Basic Wall: Exterior - Brick on Mtl. Stud	Air Barrier - Air Infiltration Barrier	502 SF
Basic Wall: Exterior - Brick on Mtl. Stud	Air Barrier - Air Infiltration Barrier	234 SF
Basic Wall: Exterior - Brick on Mtl. Stud	Air Barrier - Air Infiltration Barrier	50 SF
Air Barrier - Air Infiltration Barrier: 5		1923 SF
Basic Wall: Generic - 6"	Default Wall	95 SF
Basic Wall: Generic - 6"	Default Wall	108 SF
Basic Wall: Generic - 8"	Default Wall	1010 SF
Basic Wall: Generic - 8"	Default Wall	17 SF
Basic Wall: Generic - 8"	Default Wall	40 SF

A window opens with the new schedule.

10. Close without saving.

Exercise 8-13

Create a Mass Floor Schedule

Drawing Name: **mass_schedule.rvt**
Estimated Time to Completion: 10 Minutes

Scope
Create a mass floor schedule

Solution

1. Activate the **View** ribbon.
   Select **Schedules→Schedule/ Quantities** from the Create panel.

2. 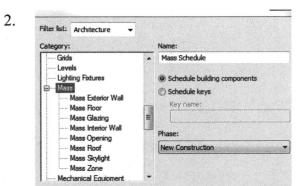 Highlight **Mass**.
   Press **OK**.

3. Add the following fields:
   - **Family and Type**
   - **Gross Floor Area**
   - **Gross Volume**
   - **Description**

4. 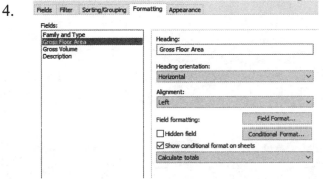 Select the Formatting tab.
   Select the **Gross Floor Area**.
   Select **Calculate totals** from drop-down list.

5.

Graphics

Build schedule: ◉ Top-down
○ Bottom-up

Grid lines: ☑ Thin Lines ▾   ☐ Grid in headers/footers/spacers

Outline: ☐ Thin Lines ▾

Height: Variable ▾   ☐ Blank row before data

Select the Appearance tab.
Disable **Blank row before data**.
Press **OK**.

6.

\<Mass Schedule\>

A	B	C	D
Family and Type	Gross Floor Area	Gross Volume	Description
Building 1: Building 1	40724 SF	475111.08 CF	
Building 2: Building 2	4973 SF	58017.48 CF	
Building 3: Building 3	59395 SF	695765.31 CF	
Building 4: Building 4	66309 SF	779134.58 CF	
Towers: Towers	28104 SF	331007.33 CF	
Building 5: Building 5	15398 SF	174509.62 CF	

7. Close the file without saving.

Exercise 8-14
Create a Schedule Template

Drawing Name: **schedule templates.rvt**
Estimated Time to Completion: 20 Minutes

Scope
Create a view template for a schedule to re-use in other projects.
Use Transfer Project Standards to import view templates into a new project
Assign a view template to a schedule

Solution

1. 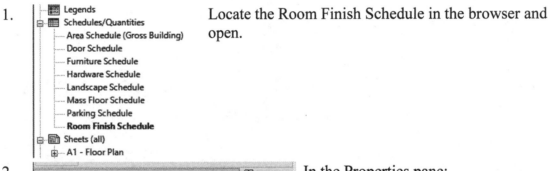 Locate the Room Finish Schedule in the browser and open.

2. In the Properties pane:
Select **<None>** next to View Template.

3. 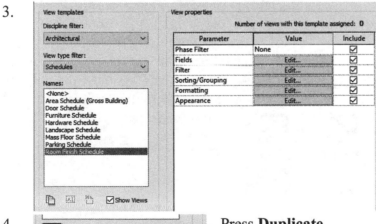 Enable **Show Views**.
Select **Architectural**.
Select **Schedules** under View type filter:
Highlight **Room Finish Schedule**.

4. Press **Duplicate**.

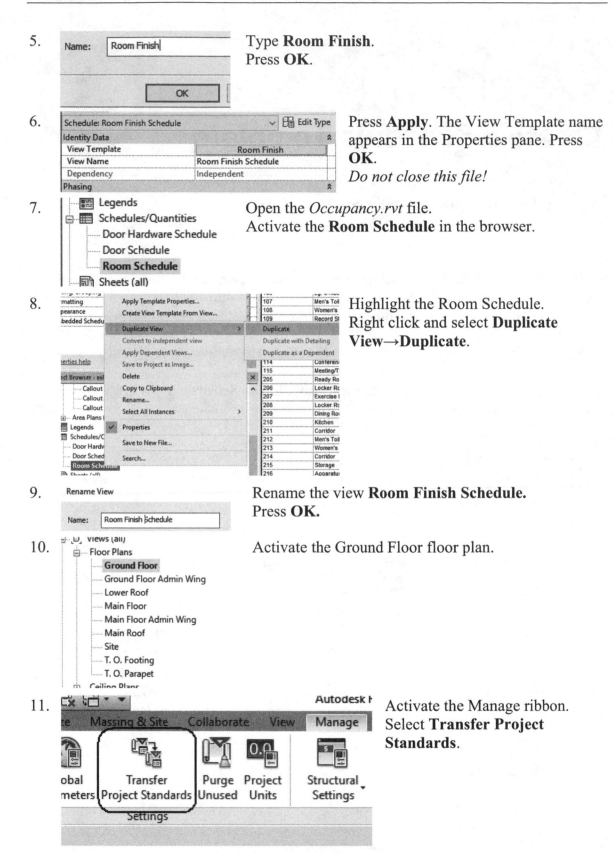

5. Type **Room Finish**.
Press **OK**.

6. Press **Apply**. The View Template name appears in the Properties pane. Press **OK**.
Do not close this file!

7. Open the *Occupancy.rvt* file.
Activate the **Room Schedule** in the browser.

8. Highlight the Room Schedule.
Right click and select **Duplicate View→Duplicate**.

9. Rename the view **Room Finish Schedule.**
Press **OK.**

10. Activate the Ground Floor floor plan.

11. Activate the Manage ribbon.
Select **Transfer Project Standards**.

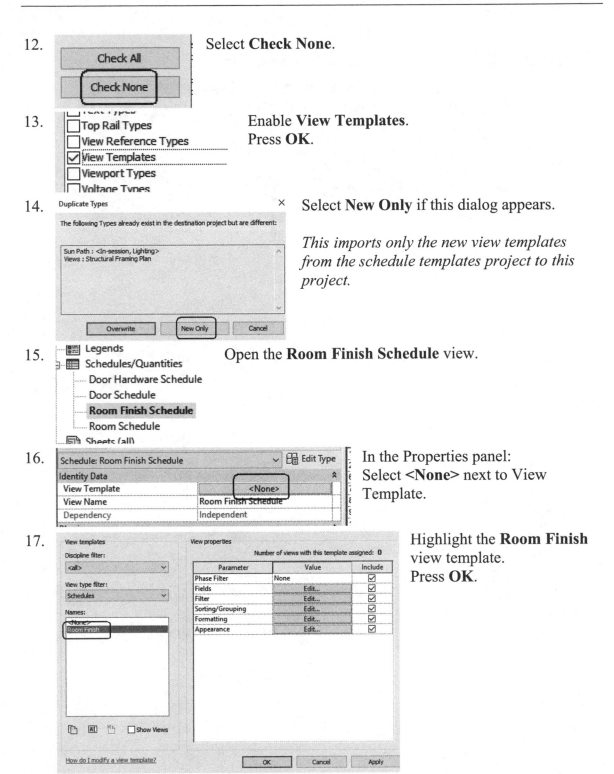

12. Select **Check None**.

13. Enable **View Templates**.
Press **OK**.

14. Select **New Only** if this dialog appears.

    This imports only the new view templates from the schedule templates project to this project.

15. Open the **Room Finish Schedule** view.

16. In the Properties panel:
Select **<None>** next to View Template.

17. Highlight the **Room Finish** view template.
Press **OK**.

18. The schedule updates with the view template's settings.

Number	Name	Floor	Wall	Ceiling	Area	Comments	Level
1	Stair #1	CONCRETE	GYPSUM PLASTER	GYPSUM PLASTER	200 SF		Ground Floor
2	Stair #2	CONCRETE	GYPSUM PLASTER	GYPSUM PLASTER	191 SF		Ground Floor
6	ROOM				230 SF		Main Floor
7	ROOM				235 SF		Main Floor
8	ROOM				231 SF		Main Floor
9	ROOM				227 SF		Main Floor
100	Vestibule	GRANITE	GYPSUM PLASTER	GYPSUM PLASTER	136 SF	Public Entrance	Ground Floor
101	Lobby	CARPET	GYPSUM PLASTER	GYPSUM PLASTER	241 SF	Public Entrance	Ground Floor
102	Corridor	VCT	GYPSUM PLASTER	GYPSUM PLASTER	410 SF		Ground Floor
103	Office	VCT	GYPSUM PLASTER	GYPSUM PLASTER	143 SF		Ground Floor
104	Office	VCT	GYPSUM PLASTER	GYPSUM PLASTER	143 SF		Ground Floor
105	Office	VCT	GYPSUM PLASTER	GYPSUM PLASTER	143 SF		Ground Floor
106	Lg. Office	VCT	GYPSUM PLASTER	GYPSUM PLASTER	183 SF		Ground Floor
107	Men's Toilet Room	VCT	GYPSUM PLASTER	GYPSUM PLASTER	108 SF		Ground Floor
108	Women's Toilet Ro	VCT	GYPSUM PLASTER	GYPSUM PLASTER	155 SF		Ground Floor
109	Record Storage	VCT	GYPSUM PLASTER	GYPSUM PLASTER	123 SF		Ground Floor
110	Fax/Copy Room	VCT	GYPSUM PLASTER	GYPSUM PLASTER	58 SF		Ground Floor

19. Close without saving.

Certified User Practice Exam

1. **True or False**: All schedules update automatically when you modify the project.

2. A _____ is a graphical representation of an annotation element or other object.

   A. Schedule
   B. Image
   C. Symbol
   D. Detail

3. **True or False**: You can include graphical images in schedules.

4. To add a graphical image to a schedule:

   A. You need to create a 3D camera view.
   B. You need to create a rendering.
   C. You need to create or have an image file.
   D. You need to modify the scheduled element's type properties.

5. When you add text notes to a view using the Text tool, you can control all of the following except:

   A. The display of leader lines
   B. Text Wrapping
   C. Text Font
   D. Whether the text is bold or underlined

6. To use a different text font, you need to:

   A. Use the Text tool.
   B. Create a new Text family and change the Type Properties.
   C. Create a new Text family and change the Instance Properties.
   D. Modify the text.

Answers
1) True; 2) C; 3) True; 4) C; 5) C; 6) B

Certified Professional Practice Exam

1. Legend views:

    A. Are placed on only one sheet at a time
    B. Contain model and annotation elements
    C. Have model components tagged
    D. Have dimensions added to system families

2. To add or remove fields from a revision schedule:

    A. Select the revision schedule in the browser, right click and select Properties
    B. Select the Schedules tool on the View ribbon.
    C. First you must open the title block family for editing
    D. Select the Revision Table in the graphics window, right click and select Element Properties

3. To add revisions to the Sheet Issues/Revisions dialog, you must access this ribbon:

    A. View
    B. Manage
    C. Annotate
    D. Insert

4. On which tab of the Schedule Properties dialog can hidden fields be enabled?

    A. Fields
    B. Filter
    C. Formatting
    D. Appearance
    E. Sorting/Grouping

5. Color Schemes can be placed in the following views (Select 3):

    A. Floor plan
    B. Ceiling Plan
    C. Section
    D. Elevation
    E. 3D

6. To define the colors and fill patterns used in a color scheme legend, select the legend, and Click Edit Scheme on the:

    A. Properties palette
    B. View Control Bar
    C. Ribbon
    D. Options Bar

Room Schedule			
Number	Name	Area	Level
4	Accounting	49.29	Level 1
3	CEO	117.84	Level 1
9	Common Area	391.84	Level 1
6	Conference Room	146.21	Level 1
7	Copy/Mail Room	71.73	Level 1
1	Engineering	136.75	Level 1
8	Lobby	236.81	Level 1
5	Operations	49.29	Level 1
2	Sales/Marketing	119.02	Level 1
Grand total: 9		1318.77	

7. The Room Schedule shown sorts by Level and by _____.

    A. Name
    B. Area
    C. Number
    D. Total

8. A legend is a view that can be placed on:

    A. A plan view
    B. A drafting view
    C. Multiple Plan Regions
    D. Multiple Sheets

9. The calculated total for a mass floor area can be displayed in:

    A. Project Browser
    B. Properties pane
    C. Mass Floor Schedule
    D. Floors Schedule

10. To compute the total cost of a material in a schedule, you need to create a: _____.

    A. Project parameter for cost
    B. Shared Parameter
    C. Calculated Value
    D. Project Filter

11. A _____ is a view that displays information about a building project in tabular form.

    A. Schedule
    B. Graph
    C. Legend
    D. Chart

12. Schedules can be filtered by the following disciplines EXCEPT:

    A. Architecture
    B. Structure
    C. Mechanical
    D. Electrical
    E. Piping
    F. Civil

13. To change the font used in a Color Scheme Legend:

    A. Select the legend, then select Edit Scheme from the ribbon.
    B. Select the text in the legend, double click to edit.
    C. Select the legend, then select Edit Type from the Properties pane.
    D. Select the legend, right click and select Edit Family.

Answers
  1) B; 2) C; 3) A; 4) C; 5) A, C, D; 6) C; 7) A; 8) D; 9) C; 10) C; 11) A; 12) F; 13) C

Presenting the Building Model

This lesson addresses the following certification exam questions:

- Sun and Shadow Settings
- Rendering
- Decals
- Hidden Views
- Graphic Display Overrides

Revit Architecture can create photorealistic images of both exterior and interior views of your model. Users can create lighting, plants, decals, and place people in their model.

If you need to monitor how much memory the rendering process is using in the Windows Task Manager, the rendering process is named fbxooprender.exe. When you render an image, the rendering process may use up to 4 CPUs.

Because rendering can use up system resources, turn off active screen savers, and shut down any non-essential processes. (For example, close your email program and don't browse the internet during rendering.) By closing applications, more CPU capacity will be made available to the rendering process and can reduce render time. Many users will have two workstations: one for rendering and one for regular office work if they create a lot of renderings.

If you experience long delays when rendering, use the Windows Task Manager to monitor processes. If fbxooprender.exe is not using close to 99% of processor power, other active processes may be interfering with the rendering process. Shut down non-essential tasks to make more processor power available for the rendering process.

Autodesk now allows users the option of rendering on their cloud server. This allows you to keep working during the rendering process.

Exercise 9-1
Creating a Toposurface

Drawing Name: **C_Condo_complex.rvt**
Estimated Time to Completion: 40 Minutes

Scope
Create a toposurface.
Add site components.
Add entourage.

Solution

1. Floor Plans
    Level 1
    Level 2
    Level 3
    Roof
    Site

    Activate the **Site** view.
    Turn off the visibility of grids, elevations, and sections.

2. Massing & Site | Coll
    Wall Floor | Toposurface

    Activate the **Massing & Site** ribbon.
    Select the **Toposurface** tool.

3. Place Point

    Use the **Place Point** tool to create an outline of a lawn surface.

4.

    Pick the points indicated to create a lawn expanse.
    You can grab the points and drag to move into the correct position.

5. Click on the **Modify** tool on the ribbon to exit out of the add points mode.

6. Left click in the **Material** column on the Properties pane.

7. In the search field, type site.
Site materials will then be listed.
Highlight **Site - Grass**.

8. On the Graphics tab:
Enable **Use Render Appearance.**
Press **OK** to close the Material Browser dialog.

9. You should see **Site-Grass** in the Properties pane.

10. Select the **Green Check** on the Surface panel to **Finish Surface.**

11. Switch to **Realistic** to see the grass material.

12.

Wireframe
Hidden Line
Shaded
Consistent Colors
Realistic
Ray Trace

Switch the display to **Hidden Line** to make it easier to place the building pad.

13.

Massing & Site   Coll

Wall   Floor   Toposurface

Face

Activate the **Massing & Site** ribbon.
Select the **Toposurface** tool.

14.

Place Point

Use the **Place Point** tool to create an outline of a lawn surface.

15.

Select the four corners of the building to form a rough rectangle.

16.

Modify

Select ▾

Click on the **Modify** tool on the ribbon to exit out of the add points mode.

17.

Topography

Materials and Finishes
Material   <By Category>

Left click in the **Material** column on the Properties pane.

18.

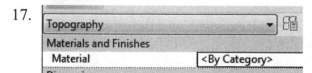

con

Project Materials: All ▾
Search results for "con"

Name

Concrete - Cast-in-Place Concrete

In the search field, type **concrete**.
Highlight **Concrete - Cast In-Place Concrete**.

19. Enable **Use Render Appearance** in the Material Editor dialog.
Press **OK** to close the Material Browser dialog.

20. You should see **Concrete: Cast-In-Place Concrete** in the Properties pane.

21. Press **OK**.

22. Select the **Green Check** on the Surface panel to **Finish Surface**.

23. Select the **Building Pad** tool from the Model Site panel.

24. Select the **Rectangle** tool from the Draw panel

25. Use **Rectangle** to create a sidewalk up to the left entrance of the building.

The rectangle must be entirely within the toposurface or you will get an error message.

26. On the Properties pane: Select **Edit Type**.

27. Select **Duplicate**.

28. Enter **Walkway** in the Name field.
Press **OK**.

29. Select **Edit** under Structure.

30. Assign the Masonry- Brick material to the structure layer.

31. Select Masonry-Brick as the material.

32. Enable **Use Render Appearance for Shading**.
Press **OK** to exit all the material dialogs.

33. In the Properties pane:
Set the Height Offset from Level to **1″**.

This protrudes the walkway 1″ above the toposurface so it appears better.

34. Select the **Green Check** on the Mode panel to **Finish Building Pad**.

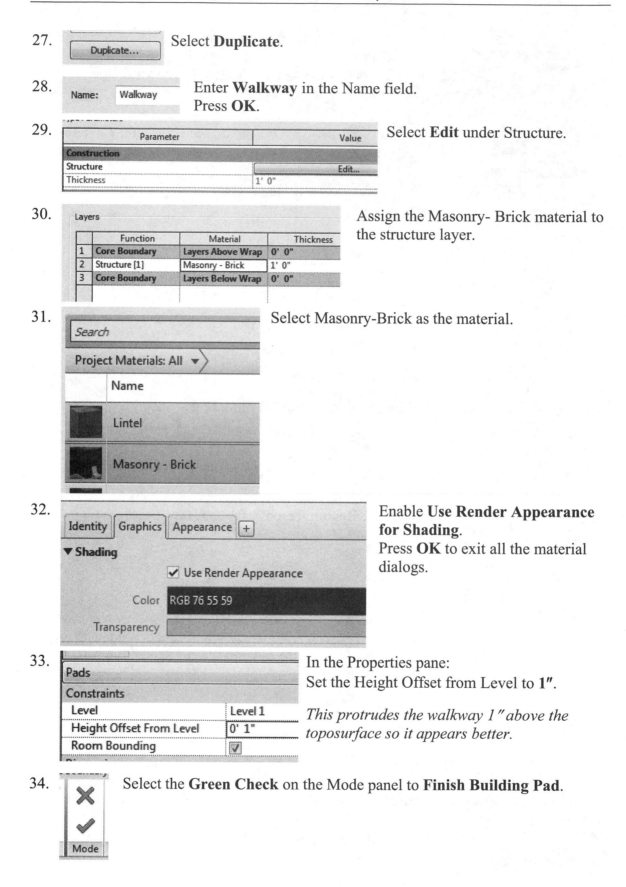

35. 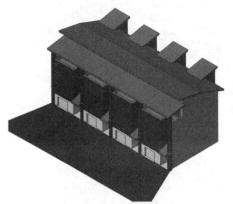 Switch to a **3D** view.

36. Use View Properties (VP) to set the style to **Shaded**.
Rotate the view so you can see the topo surface.

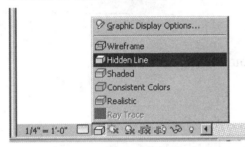

37. Switch to the **Site** view.

38. Switch the display to **Hidden Line**.

39. Activate the **Massing & Site** ribbon.

40. Select the **Site Component** tool on the Model Site panel.

41.  Select the **Load Family** tool from the Mode panel.

42. Select the *Planting* folder.

43. 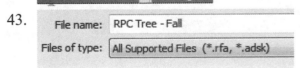 Select the *RPC Tree – Fall.rfa*.
Press **Open**.

44.  Select **Japanese Maple – 10′** from the Type Selector drop-down list.

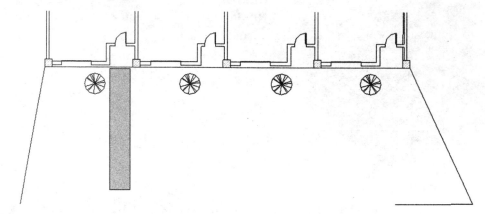

45. Place a tree at each entrance.

46. 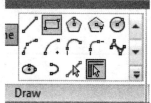 Select the **Building Pad** tool from the Model Site panel.

47. Select the **Rectangle** tool from the Draw panel

48.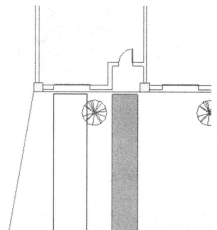

Use **Rectangle** to create a drive up to the garage door of the building.

The rectangle must be entirely within the toposurface or you will get an error message.

49. 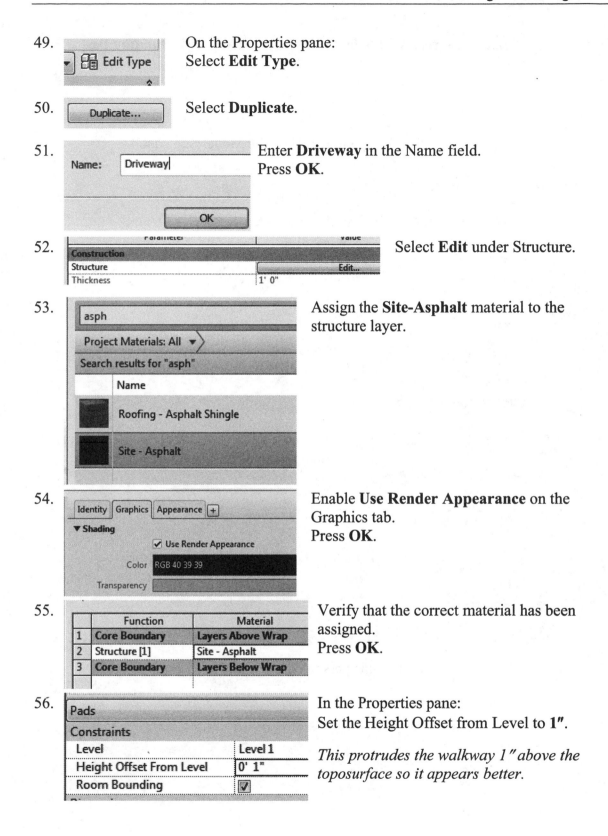 On the Properties pane:
Select **Edit Type**.

50. Select **Duplicate**.

51. Enter **Driveway** in the Name field.
Press **OK**.

52. Select **Edit** under Structure.

53. Assign the **Site-Asphalt** material to the structure layer.

54. Enable **Use Render Appearance** on the Graphics tab.
Press **OK**.

55. Verify that the correct material has been assigned.
Press **OK**.

56. In the Properties pane:
Set the Height Offset from Level to **1"**.

This protrudes the walkway 1" above the toposurface so it appears better.

57.   Select the **Green Check** on the Mode panel to **Finish Building Pad**.

58.   Select the **Site Component** tool on the Model Site panel.

59.   Select the **Load Family** tool from the Mode panel.

60.   Browse to the *Entourage* folder.

61.   Locate the **RPC Female [M_RPC Female.rfa]** file. Press **Open**.

62.   Set the Female to **Cathy** in the Properties pane.

63.   Place the person on the walkway.

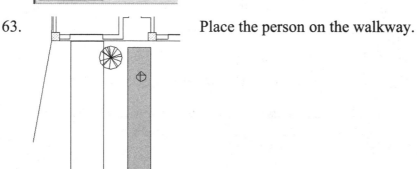

64. 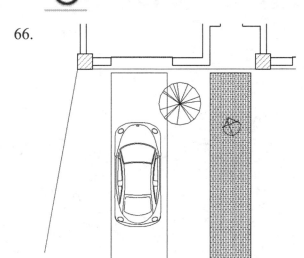 If you zoom in on the person, you only see a symbol – you don't see the real person until you perform a Render. You may need to switch to Hidden Line view to see the symbol for the person.

The point indicates the direction the person is facing.

65. Rotate your person so she is facing the building.

66. Use the site component tool to place a VW beetle on the driveway.

You will need to load the family from the entourage folder.

67. Save the file as *ex9-1.rvt*.

Tips & Tricks

Make sure **Level 1** or **Site** is active, or your trees could be placed on Level 2 (and be elevated in the air). If you mistakenly placed your trees on the wrong level, you can pick the trees, right click, select Properties, and change the level.

Exercise 9-2

Defining Camera Views

Drawing Name: **ex9-1.rvt**
Estimated Time to Completion: 30 Minutes

Scope
Create a camera view.
Rename the view.
Duplicate the view
Set view properties.
Compare different view display settings

Solution

1. Floor Plans — Level 1 — Level 2 — Level 3 — Roof — **Site**   Activate the **Site** floor plan.

2. [View Manage Modify ribbon — Render Gallery, 3D View, Section, Callout — Default 3D View, Camera]   Activate the **View** ribbon.
    Select the **3D View→Camera** tool.

3. Click to place the eye position at the cursor location.   Your cursor icon changes to a camera.
    In the lower left prompt area: the prompt says
    "Click to place the eye position at the cursor
    location."

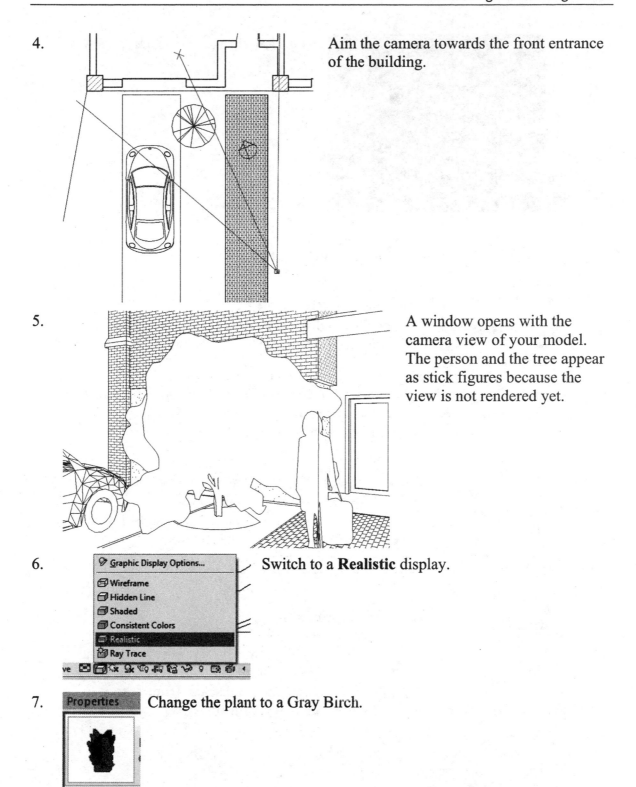

4.   Aim the camera towards the front entrance of the building.

5.   A window opens with the camera view of your model. The person and the tree appear as stick figures because the view is not rendered yet.

6.   Switch to a **Realistic** display.

7.   Change the plant to a Gray Birch.

8.  Our view changes to a colored view.

9. Ceiling Plans
   3D Views
   3D View 1
   {3D}

   If you look in the browser, you see that a view has been added to the 3D Views list.

10. Name: 3D Realistic

    Highlight the **3D View 1**.
    Right click and select **Rename**.
    Rename to **3D Realistic**.

The view we have is a perspective view – not an isometric view.
Isometrics are true scale drawings. The Camera View is a perspective view with vanishing points. Isometric views have no vanishing points. Vanishing points are the points at which two parallel lines appear to meet in perspective.

11.

Crop View	☑
Camera	
Rendering Settings	Edit...
Locked Orientation	☐
Perspective	☑
Eye Elevation	4' 6"
Target Elevation	5' 10"
Camera Position	Explicit

    In the Properties pane:
    Change the Eye Elevation to **4′6″** [**1370 mm**].
    Change the Target Elevation to **5′ 10″** [**3200 mm**].

12. Apply   Press **Apply**. Your view shifts slightly.

13.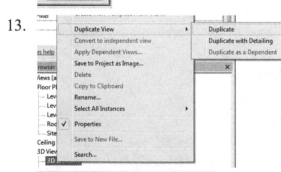

    Highlight the 3D Realistic view in the Project Browser.
    Right click and select **Duplicate View→Duplicate**.

14. Name: 3D Realistic with Shadows

    OK   Cancel

    Highlight the **3D Realistic Copy 1**.
    Right click and select **Rename**.
    Rename to **3D Realistic with Shadows**.

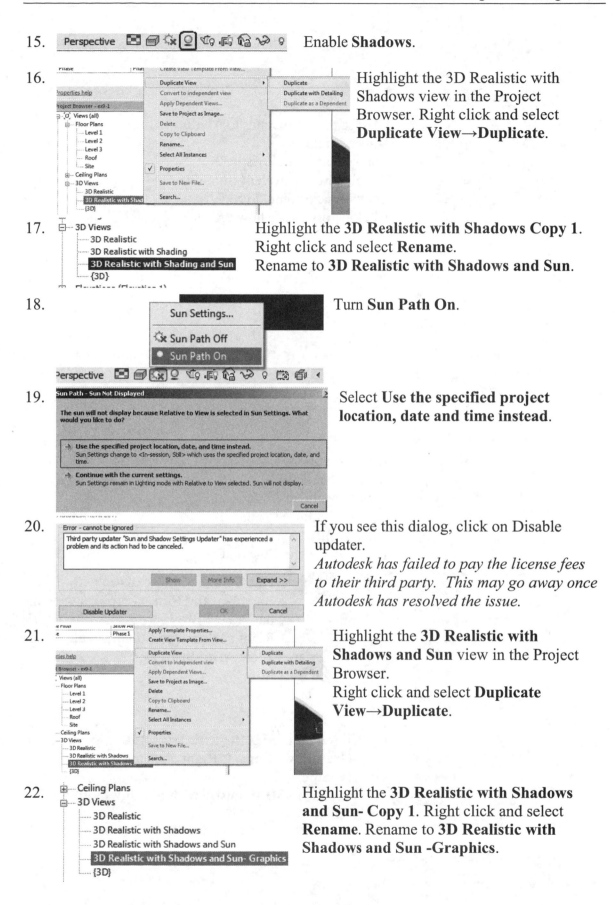

15. Perspective ▣ ▤ ◿ 🔘 ⚙ 🔲 🏠 🕊 🔘   Enable **Shadows**.

16. Highlight the 3D Realistic with Shadows view in the Project Browser. Right click and select **Duplicate View→Duplicate**.

17. Highlight the **3D Realistic with Shadows Copy 1**. Right click and select **Rename**. Rename to **3D Realistic with Shadows and Sun**.

18. Turn **Sun Path On**.

19. Select **Use the specified project location, date and time instead**.

20. If you see this dialog, click on Disable updater.
Autodesk has failed to pay the license fees to their third party. This may go away once Autodesk has resolved the issue.

21. Highlight the **3D Realistic with Shadows and Sun** view in the Project Browser.
Right click and select **Duplicate View→Duplicate**.

22. Highlight the **3D Realistic with Shadows and Sun- Copy 1**. Right click and select **Rename**. Rename to **3D Realistic with Shadows and Sun -Graphics**.

23. Click **Edit** next to Graphic Display Options.

24. Change the Silhouettes to **Medium Lines**.
Under Shadows:
Enable **Cast Shadows**.
Enable **Show Ambient Shadows**.
Set the Background to **Sky**.
Click **Apply** to see how the view changes.

25. Under Lighting:
Set the Sun to **30**.
Set the Ambient Light to **36**.
Set the Shadows to **50.**
Click **Apply** to see how the view changes.
Press **OK** to close the dialog.

26. Highlight the **3D Realistic with Shadows and Sun -Graphics** view in the Project Browser.
Right click and select **Duplicate View→Duplicate**.

27. Highlight the **3D Realistic with Shadows and Sun -Graphics – Copy 1**. Right click and select **Rename**.
Rename to **3D Realistic with Shadows and Sun - Photo**.

28. Click **Edit** next to Graphic Display Options.

29. Under Photographic Exposure:
Check **Enabled**. Enable **Manual**. Set the value to
16. Click the **Color Correction** button.

30. Set the Highlights to **0.35**.
Set the Shadows to **0.50**
Set the Saturation to **2.50**.
Set the White point to **6587**.
Press **OK** twice to exit the dialogs.

31. Highlight **Sheets** in the Project Browser.
Right click and select **New Sheet.**

32. Highlight the **E1** titleblock.
Press **OK**.

33. Rename the sheet - **Camera Views-Exterior**.

34. Drag and drop the 3D camera views you created on
to the sheet.

35. Compare the different views and how the settings affect the displays.

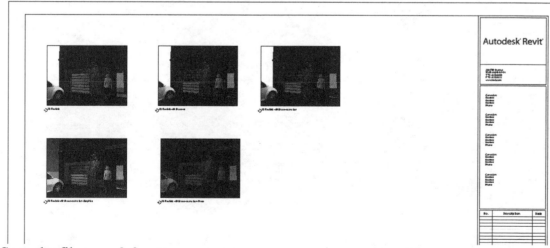

36. Save the file as *ex9-2.rvt*.

Additional material libraries can be added to Revit. You may store your material libraries on a server to allow multiple users access to the same libraries. You must add a path under Settings → Options to point to the location of your material libraries. There are many online sources for photorealistic materials; www.accustudio.com is a good place to start.

Exercise 9-3

Sun Settings

Drawing Name: **ex9-2.rvt**
Estimated Time to Completion: 15 Minutes

Scope
Sun Settings

Solution

1. Activate the **3D Realistic with Shadows and Sun** view.

3. On the View Display bar:
    Select **Sun Settings**.

4. Enable **Still** under Solar Study.

5. Select the browse button next to Location.

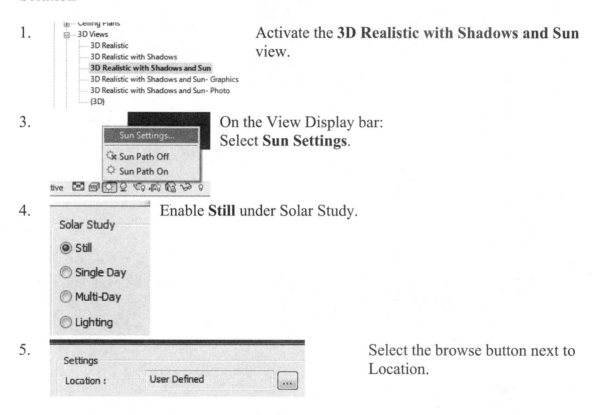

6.

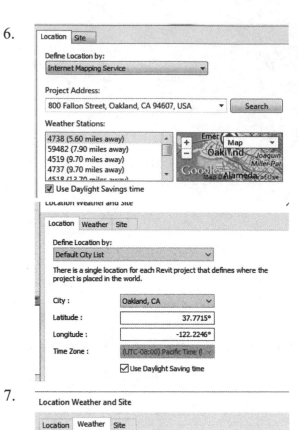

Enter an address into the Project Address field.

You must be connected to the Internet in order to use the Internet Mapping Service. If you don't have an internet connection, use the Default City List and select the city closest to your location.

Press **Search** to locate the address.
Type in the address shown.
Enable **Use Daylight Savings time**.
Press **OK**.

7.

Select the Weather tab.

Revit uses the closest weather station data for temperatures which can be used to analyze energy consumption.

Location Weather and Site

☑ Use closest weather station (OAKLAND METROPOLITAN ARPT)

Cooling Design Temperatures

	Jan	Feb	Mar	Apr	May
Dry Bulb	63 °F	67 °F	70 °F	75 °F	79 °F
Wet Bulb	58 °F	60 °F	61 °F	63 °F	66 °F
Mean Daily Range	12 °F	12 °F	12 °F	14 °F	13 °F

Heating Design Temperature: 38 °F

Clearness Number: 1.0

8.

☑ Ground Plane at Level :

Level 1

Save Settings

Enable **Ground Plane at Level: Level 1**.
Press **OK**.

9.

⊞ Ceiling Plans
⊟ 3D Views
    ⋯ 3D Realistic
    ⋯ 3D Realistic with Shadows
    ⋯ 3D Realistic with Shadows and Sun
    ⋯ 3D Realistic with Shadows and Sun- Graphics
    ⋯ 3D Realistic with Shadows and Sun- Photo
    ⋯ {3D}

Activate the 3D view.

10. Set **Sun Path On**.

11.   Select **Use the specified project location, date, and time instead**.

> Use the specified project location, date, and time instead.
> Sun Settings change to <In-session, Still> which uses the specified project location, date, and time.
>
> Continue with the current settings.
> Sun Settings remain in Lighting mode with Relative to View selected. Sun will not display.

12. 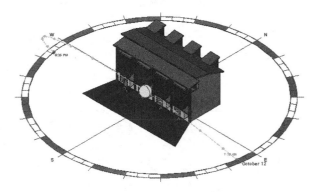 Zoom out to see the sun path. Place the cursor over the sun. Hold down the left mouse button and move the sun.

13. 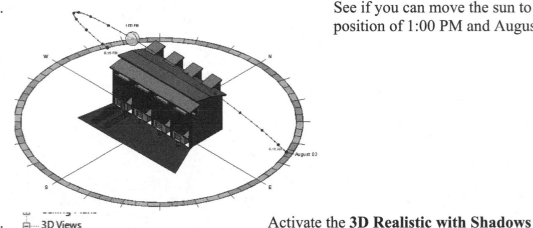 See if you can move the sun to a position of 1:00 PM and August.

14. Activate the **3D Realistic with Shadows and Sun** view.

- 3D Views
  - 3D Realistic
  - 3D Realistic with Shadows
  - **3D Realistic with Shadows and Sun**
  - 3D Realistic with Shadows and Sun- Graphics
  - 3D Realistic with Shadows and Sun- Photo

15.  On the View Display bar: Select **Sun Settings**.

16.

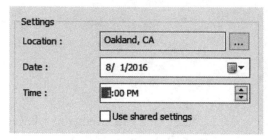

Note that the sun settings for the view are specific to each view.
Change the date to **August** and the time to **1:00 PM**.
Press **Apply** and **OK**.

17. Save as *ex9-3.rvt*.

Exercise 9-4
Rendering Settings

Drawing Name: **ex9-3.rvt**
Estimated Time to Completion: 15 Minutes

Scope
Create a rendering.
Control rendering options.
Save a rendering to the project.

Solution

1.

⊞ Ceiling Plans
☐ 3D Views
— 3D Realistic
— 3D Realistic with Shadows
— **3D Realistic with Shadows and Sun**
— 3D Realistic with Shadows and Sun- Graphics
— 3D Realistic with Shadows and Sun- Photo
— {3D}

Activate the **3D Realistic with Shadows and Sun** view.

2.

Camera	
Rendering Settings	Edit...
Locked Orientation	☐
Perspective	✓
Eye Elevation	4' 6"
Target Elevation	5' 10"
Camera Position	Explicit

On the Properties pane:
Under Camera:
Select **Edit** for Rendering Settings.

3.

Quality
Setting: Medium

Lighting
Scheme: Exterior: Sun only
Sun Settings: <In-session, Still> ...
Artificial Lights...

Background
Style: Sky: Few Clouds

Clear   Hazy
Haze: ▯

Image
Adjust Exposure...

Set the Quality Setting to **Medium**.
Press **OK**.

4.

Select the **Rendering** tool located on the Display Control bar.

5. Render — Select the **Render** button.

6.  Your window will be rendered.

7. Save to Project... — Select **Save to Project**.

8. Rendered images are saved in the Renderings branch of Project Browser.
    Name: 3D Realistic with Shadows and Sun_Rendering 1
    OK   Cancel
    Press add **Rendering 1** to the default name. Press **OK**.

9. {3D}
    Elevations (Elevation 1)
    Renderings
    3D Realistic with Shadows and Sun_Rendering 1

    Under Renderings, we now have a view called **3D Realistic with Shadows and Sun _Rendering_1**.

10. Display
    Show the model
    Select the **Show the model** button. Our window changes to – not Rendered – mode. Close the Rendering dialog.

11. Schedules/Quantities
    Sheets (all)
    B5.06 - Camera Views - Exterior
    Families
    Activate the sheet with the Camera Views.

12. Add the new rendering view to the sheet.

13. Save the file as *ex9-4.rvt*.

Exercise 9-5
Using Element Graphic Overrides in a Hidden View

Drawing Name: **hidden_views.rvt**
Estimated Time to Completion: 10 Minutes

Scope
Apply Graphic Overrides to Enhance a Hidden View

Solution

1.
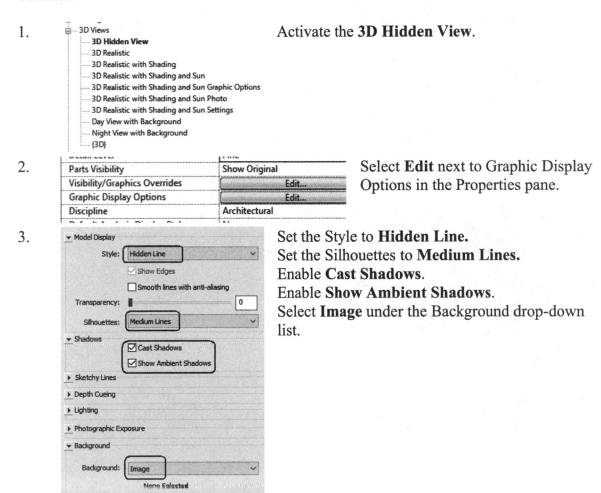
Activate the **3D Hidden View**.

   - 3D Views
     - **3D Hidden View**
     - 3D Realistic
     - 3D Realistic with Shading
     - 3D Realistic with Shading and Sun
     - 3D Realistic with Shading and Sun Graphic Options
     - 3D Realistic with Shading and Sun Photo
     - 3D Realistic with Shading and Sun Settings
     - Day View with Background
     - Night View with Background
     - {3D}

2.

Parts Visibility	Show Original
Visibility/Graphics Overrides	Edit...
Graphic Display Options	Edit...
Discipline	Architectural

Select **Edit** next to Graphic Display Options in the Properties pane.

3.

- Model Display
  - Style: Hidden Line
  - ☑ Show Edges
  - ☐ Smooth lines with anti-aliasing
  - Transparency: 0
  - Silhouettes: Medium Lines
- Shadows
  - ☑ Cast Shadows
  - ☑ Show Ambient Shadows
- Sketchy Lines
- Depth Cueing
- Lighting
- Photographic Exposure
- Background
  - Background: Image
    - None Selected

Set the Style to **Hidden Line.**
Set the Silhouettes to **Medium Lines.**
Enable **Cast Shadows**.
Enable **Show Ambient Shadows**.
Select **Image** under the Background drop-down list.

4. 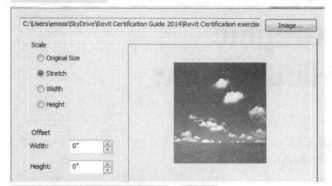 Select **Customize Image**.

5. Select the **Image** button to browse for the image file.

6.  Select *sky4.jpg*.
Press **Open**.
Press **OK**.

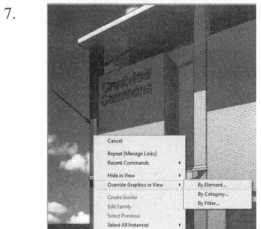 Close the Graphics Display Options by pressing **OK**.
The image is displayed in the view.

7. Select the far left column.
Right click and select **Override Graphics in View→By Element.**

8.
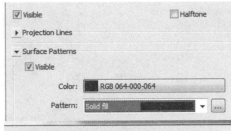

Under Surface Patterns:
Set the Color to the Color Swatch indicated.
Set the Pattern to **Solid Fill**.
Press **OK**.
Left click in the window to release the selection.

Basic colors:
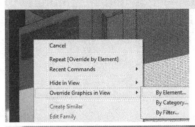

9.

Select the model text on the building.
Right click and select **Override Graphics in View→By Element.**

10.

Under Surface Patterns:
Set the Color to **Blue.**
Set the Pattern to **Solid Fill**.
Press **OK**.

11.

Try changing other elements to see what effects you can create.
Save as *ex9-10.rvt*.

Exercise 9-6
Render in Cloud

Drawing Name: **ex9-4.rvt**
Estimated Time to Completion: 10 Minutes

Scope
Render in Cloud

In order to use the Cloud, you must have an internet connection and an Autodesk 360 account. Students and teachers qualify for a free Autodesk 360 account. If you have an Autodesk subscription, you also qualify for a free account. Check with your local reseller if you are not sure about a subscription.

Solution

1.

Activate the **3D Realistic with Shadows** view.

2.

Activate the View ribbon.
Select **Render in Cloud**.

3.

Press **Continue**.

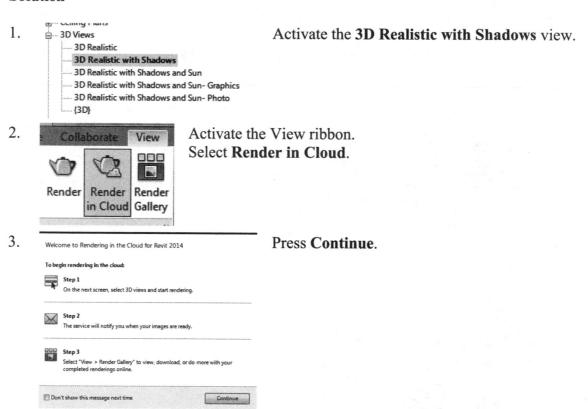

4.

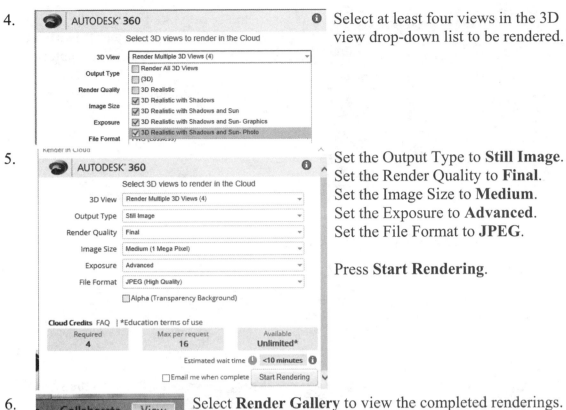

Select at least four views in the 3D view drop-down list to be rendered.

5.

Set the Output Type to **Still Image**.
Set the Render Quality to **Final**.
Set the Image Size to **Medium**.
Set the Exposure to **Advanced**.
Set the File Format to **JPEG**.

Press **Start Rendering**.

6.

Select **Render Gallery** to view the completed renderings.

7.

You should see a list of images to be reviewed.

8.  Select the image you like best.
Left click on the image and select **Download Image** to save the image to your computer.

The image will most likely be saved to the Downloads folder.

9.  Activate the **Camera Views** sheet.

10. Activate the Insert ribbon.
Select **Image.**

11. Locate the downloaded rendering file and select.

12.    Place on the sheet. Compare the image created on the Cloud against the other images.

13. Save as *ex9-5.rvt.*

Exercise 9-7
Place a Decal

Drawing Name: **ex9-5.rvt**
Estimated Time to Completion: 30 Minutes

Scope
Create a decal type.
Place a decal.
Add model text.
Join Geometry
Render.

Solution

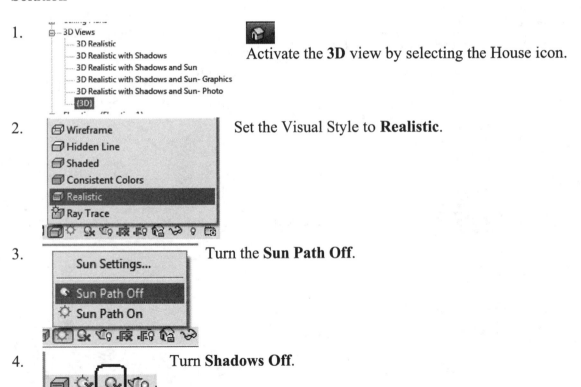

1. Activate the **3D** view by selecting the House icon.

2. Set the Visual Style to **Realistic**.

3. Turn the **Sun Path Off**.

4. Turn **Shadows Off**.

5. Activate the **Insert** ribbon.
Select **Decal Types** on the Link panel.

6. Select **Create New Decal Type**.
This tool is located on the lower left of the dialog.

7. Enter **Parking Sign** in the Name field.
Press **OK**.

Name: Parking Sign

8. Select the Browse button to select the image file to be used.

Source

Invalid Image File

9. Locate the *parking sign.png* file.
You can use a different image file if you prefer.
Press **Open**.

File name: parking sign

Files of type: All Image Files (*.bmp, *.jpg, *.jpeg, *.png, *.tif)

10. You will see a preview of the image file.
Press **OK**.

Source parking sign.png

RESIDENT PARKING UNAUTHORIZED CARS WILL BE TOWED AWAY

11. Select the **Place Decal** tool from the Link panel

12.  Place the decal on the side of the building.

13. 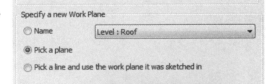 Select the Modify tool and then pick the decal that was placed.

14.  Use the grips located on the corners to enlarge the decal.

The grips can also be used to move the decal into the desired position.

15. Set   Activate the **Architecture** ribbon.
Select the **Set Work plane** tool from the Work Plane panel.

16. Enable **Pick a plane**.
Press **OK**.

Specify a new Work Plane
- ○ Name   Level : Roof
- ◉ Pick a plane
- ○ Pick a line and use the work plane it was sketched in

17. 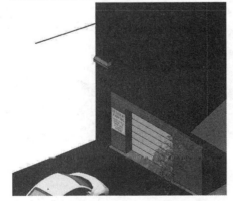 Select the wall above where the decal is placed.
If you wish to verify that you have selected the correct work plane, use the Show Work Plane tool.

18. Select the **Model Text** tool from the Model panel on the Architecture ribbon.

19. **Edit Text**

    Crestview
    Commons

    Enter **Crestview Commons** into the Edit Text dialog. *Type it as two lines as shown.*

    Press **OK**.

20.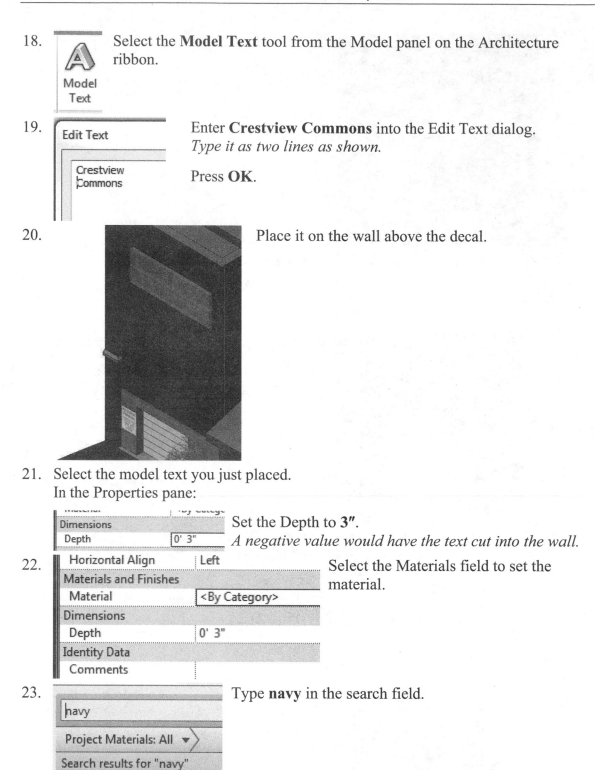

    Place it on the wall above the decal.

21. Select the model text you just placed.
    In the Properties pane:

Dimensions	
Depth	0' 3"

    Set the Depth to **3"**.
    A negative value would have the text cut into the wall.

22. | Horizontal Align | Left |
    |---|---|
    | **Materials and Finishes** | |
    | Material | <By Category> |
    | **Dimensions** | |
    | Depth | 0' 3" |
    | **Identity Data** | |
    | Comments | |

    Select the Materials field to set the material.

23. | navy |
    |---|
    | Project Materials: All ▼ |
    | Search results for "navy" |

    Type **navy** in the search field.

24. 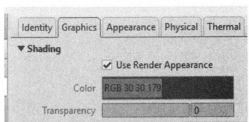 A Laminate, Navy material will be listed in the lower pane.
Click on the **Add material to document** icon.

25. 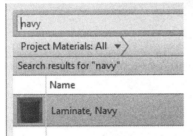 On the Graphics tab:
Enable **Use Render Appearance**.

26. 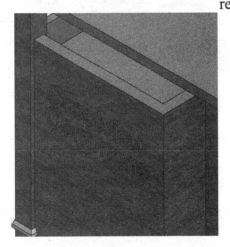 Highlight the material and press **OK**.

27. With the Model Text selected:
Set the Horizontal Align to **Center** in the Properties pane. Left click in the window to release the selection.

28. Save the file as *ex9-6.rvt*.

Exercise 9-8

Adding a Background

Drawing Name: **background.rvt**
Estimated Time to Completion: 30 Minutes

Scope
Modify Graphic Display Options
Add a background to a view
Render a view
Compare realistic and rendered views

Solution

1.
```
3D Views
    3D Realistic
    3D Realistic with Shadows
    3D Realistic with Shadows and Sun
    3D Realistic with Shadows and Sun- Graphics
    3D Realistic with Shadows and Sun- Photo
    Day View with Background
    Night View with Background
    {3D}
```
Activate the **Day View with Background** view.

2.
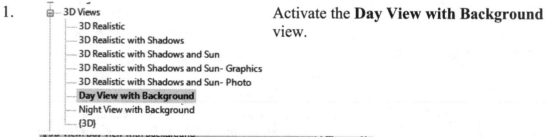

Select **Edit** next to Graphic Display Options.

3. Set the Style to **Realistic**.
Enable **Show Edges**.
Set Silhouettes to **Medium Lines**.
Enable **Cast Shadows**.
Enable **Show Ambient Shadows**.
Set Lighting to **Exterior: Sun Only**.
Set Sun to **30**.
Set Ambient Light to **40**.
Set Shadows to **50**.

4. Under Background:
Select **Image**.
Select **Customize Image**.

5. Select the **Image** button.

6. Select *sky1.png*. Press **Open**.
I have included several images you can use for backgrounds, so you can try out different backgrounds to get different effects.

7. The background will preview.
Press **OK**.

8.

Press **Apply**.
Press **OK**.
The image will update with the background.
Repeat steps 5 through 9 and use *sky2.jpeg*.

9. Verify that both sun and shadows are set to **ON**.

10. Select **Sun Settings** on the View Display bar.

Sun Settings...
☼✗ Sun Path Off
☼ Sun Path On

11. Set the Sun Settings to **8/1/2010**.
1:00 PM.
Press **OK.**

Settings
Location :   800 Fallon Street, Oakland, C [...]
Date :   8/ 1/2010 ▦▾
Time :   1:00 PM ▲▼
☐ Use shared settings

12. Select the **Render** tool on the View ribbon.
You can also select the Render tool on the display bar at the bottom of the screen.

Render   Render   Render
         in Cloud Gallery

13.

Set the Quality to **Medium**.
Set the Scheme to **Exterior: Sun Only**.
Set the Background to **Image**.
Select **Customize Image** and select *sky2.jpg* for the background.

Press **Render**.

Render ☐ Region

Quality
Setting: Medium

Output Settings
Resolution: ⦿ Screen
○ Printer

Width: 1080 pixels
Height: 810 pixels
Uncompressed image size: 3.3 MB

Lighting
Scheme: Exterior: Sun only
Sun Setting: <In-session, Still> [...]
Artificial Lights...

Background
Style: Image
Customize Image...
sky2.JPG

Image
Adjust Exposure...

14.

Select **Save to Project**.

Save to Project...

15.

Rendered images are saved in the Renderings branch of Project Browser

Name: Day View with Background

Press **OK** to accept the name. Close the rendering dialog.

16.

Activate **Night View with background**.

⊞ Ceiling Plans
⊟ 3D Views
├ 3D Realistic
├ 3D Realistic with Shadows
├ 3D Realistic with Shadows and Sun
├ 3D Realistic with Shadows and Sun- Graphics
├ 3D Realistic with Shadows and Sun- Photo
├ Day View with Background
├ **Night View with Background**
└ {3D}

17.

Verify that both sun and shadows are set to **ON**.

18.

Select **Sun Settings** on the View Display bar.

Sun Settings...
☼ Sun Path Off
☼ Sun Path On

19.

Set the Sun Settings to **8/1/2010**.
6:00 PM.

Press **OK**.

20.

Select **Edit** next to Graphic Display Options.

21.

Set the Style to **Realistic**.
Enable **Show Edges**.
Set Silhouettes to **Medium Lines**.
Enable **Smooth lines with anti-aliasing**.
Enable **Cast Shadows**.
Enable **Show Ambient Shadows**.
Set Scheme to **Exterior: Sun and Artificial**.
Set Sun to **8**.
Set Ambient Light to **30**.
Set Shadows to **50**.
Press **Apply**.

22.

Under Background:
Select **Image**.
Select **Customize Image**.

23.

Select the **Image** button.

24.

Select *night4.jpg*.
Press **OK**.

25.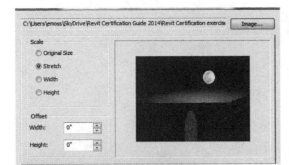

The background will preview.
Press **OK**.

26.

Press **Apply**.
The image will update with the background.
Press **OK**.

27.

Select the **Render** tool in the View Display bar.

28.

Set the Quality to **Medium**.
Set the Scheme to **Exterior: Sun and Artificial Lights**.
Set the Background to **Image**.
Select **Customize Image** and select *night4.jpg* for the background.

Press **Render**.

29.

Select **Save to Project**.

Save to Project...

30.

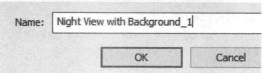

Name: Night View with Background_1

OK   Cancel

Press **OK** to accept the default name. Close the rendering dialog.

31.

Sheets (all)
   B5.06 - Camera Views - Exterior
   B5.07 - Exterior Background Views
  Families

Activate the **Exterior-Background Views** sheet.

32.

Add the realistic views and the rendering views to the sheet so you can compare the different views.

Save as *ex9-7.rvt*.

Exercise 9-9
Using Transparency Settings

Drawing Name: **office.rvt**
Estimated Time to Completion: 30 Minutes

Scope
Apply transparency to elements

Solution

1.   3D Views
       3D Section
       Inside Cubicle
       Outside Cubicle
       {3D}
                         Activate the **3D Section** view.

2. Hold down the Control key.
Select the walls for the first cubicle.

3.

                               Right click and select **Override Graphics in View→By Element.**

4.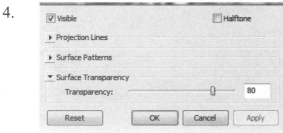

                               Set the Surface Transparency to **80**.
Press **OK**.

5.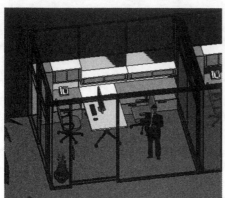

You can now see clearly inside the cubicle. Close without saving.

Exercise 9-10
Custom Render Settings

Drawing Name: **office.rvt**
Estimated Time to Completion: 15 Minutes

Scope
Create custom rendering settings

If the link for the decal needs to be reset: Select the Decal, Select Edit Type, Edit Decal Attributes and select the decal – street scene.

Solution

1. Activate the **Inside Cubicle 3D** view.

2. Select the **Render** tool from the View Control bar.

3. Under Quality:
   Setting: Select **Edit**.

4. Set the Setting to **Custom**.

5. Enable **Advanced**.
   Enable **Render by Time**.
   Set how many minutes you want to allow for the rendering.

   I entered 30 – you may want to do this exercise prior to a lunch break and set it to 30 to 60 minutes.

   Press **OK**.

6. Set the Lighting Scheme to **Interior: Sun and Artificial**.

Lighting
Scheme: Interior: Sun and Artificial ▾
Sun Setting: <In-session, Still> [...]
Artificial Lights...

7. Enable **Region**.
This renders just a small sample of the view.

Render ☑ Region

8. Select the region and use the grips to position and size the desired region for rendering.

9. Press **Render**.

Render

10. Select **Adjust Exposure**.

Image
Adjust Exposure...
Save to Project...   Export...

11. Increase the Exposure Value to 14.5 or greater.
Note how the rendering adjusts.
Press **OK**.

Exposure Control
Reset to Default
Settings
Exposure Value: 14.64   Brighter — Darker
Highlights: 0.25   Darker — Brighter
Mid Tones: 1   Darker — Brighter
Shadows: 0.2   Lighter — Darker
White Point: 6500   Cooler — Warmer
Saturation: 1   Grey — Intense

12.  Close the file without saving.

Certified User Practice Exam

1. Renderings can be created in which TWO view types?
   A. Elevation
   B. Plan
   C. Section
   D. 3D
   E. Camera

2. THREE items important for accurate shadow studies:
   A. Shared parameters
   B. Toposurface
   C. Color fill schemes
   D. Project location
   E. Sun position (date and time)

3. In the *default.rte* template, what level is the Ground Plane set in the Sun and Shadow Settings?
   A. Site
   B. Level 1
   C. Level 2

4. You can create a toposurface by placing points OR:
   A. creating from import
   B. drawing contour lines
   C. drawing a path
   D. picking points

5. In a view, you want to use the Visibility/Graphics dialog to override the graphics of an element category, but the category displays in gray, and you cannot change it. What is the problem?
   A. The visibility of the category may be controlled by a view template.
   B. The element category is locked.
   C. The view display is set to Hidden.
   D. The view is a Section.

Answers
   1) D & E; 2) B, D & E; 3) B; 4) B; 5) A; 8) D

Certified Professional Practice Exam

1. The Place Decal tool is located on which ribbon:
    A. Architecture
    B. Insert
    C. View
    D. Annotations

2. Before you add a building pad, you need a:
    A. floor
    B. toposurface
    C. wall
    D. isolated foundation

3. When creating a rendering, you can control the Exposure Control by selecting this on the Render dialog:
    A. Quality Setting
    B. Sun Setting
    C. Adjust Exposure
    D. Lighting

4. Select the display setting that allows you to preview a decal.
    A. Wireframe
    B. Hidden line
    C. Consistent Colors
    D. Realistic

5. Decals can be placed on _____ and _____ surfaces.
    A. Flat
    B. Slanted
    C. Round
    D. Visible

6. Backgrounds can be added in ALL of the following view types EXCEPT:
    A. Elevation
    B. Section
    C. Detail
    D. Isometric
    E. Camera (Perspective) 3D

7. Types of backgrounds are ALL of the following EXCEPT:
    A. Sky
    B. Gradient
    C. Image
    D. Region

Answers
   1) B; 2) B; 3) C; 4) D; 5) A & C; 6) C; 7) D

Lesson

10

Collaboration

This lesson addresses the following Professional certification exam questions:

- Demonstrate how to copy and monitor elements in a linked file
- Apply interference checking in Revit
- Using Shared Coordinates
- Worksets
- Linked files

There are no questions regarding collaboration on the user certification exam.

In most building projects, you need to collaborate with outside contractors and with other team members. A mechanical engineer uses an architect's building model to layout the HVAC (heating and air conditioning) system. Proper coordination and monitoring ensures that the mechanical layout is synchronized with the changes that the architect makes as the building develops. Effective change monitoring reduces errors and keeps a project on schedule.

Worksets are used in a team environment when you have many people working on the same project file. The project file is located on a server (a central file). Each team member downloads a copy of the project to their local machine. The person is assigned a workset consisting of building elements that they can change. If you need to change an element that belongs to another team member, you issue an Editing Request which can be granted or denied. Workers check in and check out the project, updating both the local and server versions of the file upon each check in/out.

Project sharing is the process of linking projects across disciplines. You can share a Revit Structure model with an MEP engineer.

You can link different file formats in a Revit project, including other Revit files (Revit Architecture, Revit Structure, Revit MEP), CAD formats (DWG, DXF, DGN, SAT, SKP), and DWF markup files. Linked files act similarly as external references (XREFs) in AutoCAD. You can also use file linking if you have a project which involves multiple buildings.

It is recommended to use linked Revit models for

- Separate buildings on a site or campus.

- Parts of buildings which are being designed by different design teams or designed for different drawing sets.

- Coordination across different disciplines (for example, an architectural model and a structural model).

Linked models may also be appropriate for the following situations:

- Townhouse design when there is little geometric interactivity between the townhouses.

- Repeating floors of buildings at early stages in the design, where improved Revit model performance (for example, quick change propagation) is more important than full geometric interactivity or complete detailing.

You can select a linked project and bind it within the host project. Binding converts the linked file to a group in the host project. You can also convert a model group into a link which saves the group as an external file.

I have included a portion of the worksets exercise I do in my classroom. This exercise requires that there is a shared location on a server where students have read/write access. In many classroom settings, the IT department only provides students with read access and they can only write to a local flash drive. This is to prevent file corruption and minimize exposures to computer viruses. Instructors and users should keep this in mind during the workset exercise.

Exercise 10-1
Monitoring a Linked File

Drawing Name: **i_multiple_disciplines.rvt**
Estimated Time to Completion: 40 Minutes

Scope
Link a Revit Structure file
Monitor the levels in the linked file
Reload the modified Structure file
Perform a coordination review
Create a Coordination Review report

Solution

1. Select **Options** on the Application menu.

2. In the Username field, type in your first initial and last name.
Press **OK**.
Revit will auto-fill the username in reports and the title block.

3. If you have an A360 Account, the sign in for the user account is used for your user name. If you don't want to use the name displayed by the A360 account, you need to log out of the A360 account.

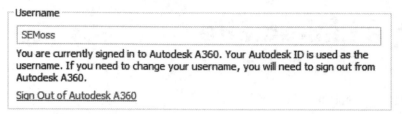

4. Open the *i_multiple_disciplines.rvt* file.

5. 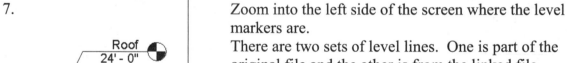 Activate the **Insert** ribbon.
    Select the **Link Revit** tool on the Link panel.

6. 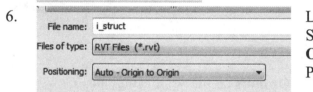 Locate the *i_struct* file.
    Set the Positioning to **Auto-Origin to Origin**.
    Press **Open**.

7. Zoom into the left side of the screen where the level markers are.
    There are two sets of level lines. One is part of the original file and the other is from the linked file.

    Roof
    24' - 0"

    Roof TOS
    23' - 4 1/4"

8. Roof
    24' - 0"
    If you mouse over the linked file it will show information.

    Roof TOS
    23' - 4 1/4"

    RVT Links : Linked Revit Model : i_struct.rvt : 4 : location <Not Shared>

9. 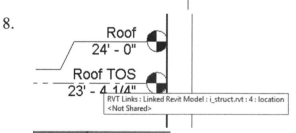 Activate the Collaborate ribbon.
    Select **Copy/Monitor→Select Link** from the Coordinate panel.
    Pick in the window to select the linked file.

    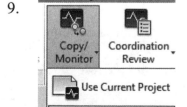

10. 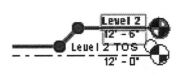 Select the **Monitor** tool from the Tools panel.

11. 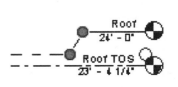 Select the Level 1 level line twice. The first selection will be the host file. The second selection will be the linked file. This is because the level lines overlap.

12. You should see a symbol on Level 1 indicating that Level 1 is currently being monitored for changes.

13. Select the Level 2 level line. The first selection will be the host file. The second selection will be the linked file. You should see a tool tip indicating a linked file is being selected.

14. Zoom out. You should see a symbol on Level 2 indicating that Level 2 is currently being monitored for changes.

15. Select the Roof level line.
The first selection will be the host file.
The second selection will be the linked file.
You should see a tool tip indicating a linked file is being selected.

16. Zoom out. You should see a symbol on the Roof level indicating that it is currently being monitored for changes.

17. Warning

Elements already monitored

If you try to select elements which have already been set to be monitored, you will see a warning dialog.
Simply close the dialog and move on.

18. Select **Finish** on the Copy/Monitor panel.

19.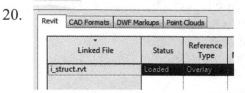

Activate the **Insert** ribbon.
Select **Manage Links** from the Link panel.

20.

Revit	CAD Formats	DWF Markups	Point Clouds		
Linked File		Status	Reference Type		
i_struct.rvt		Loaded	Overlay	

Select the **Revit** tab.
Highlight the *i_struct.rvt* file.

21. Reload From...   Select **Reload From**.

22.

File name: i_struct_revised
Files of type: RVT Files (*.rvt)

Locate the *i_struct_revised* file.
Press **Open**.

23.

Warning - can be ignored

Instance of link needs Coordination Review

Show   More Info   Expand >>

A dialog will appear indicating that the revised file requires Coordination Review.
Press **OK**.

24.

Revit	CAD Formats	DWF Markups	Point Clouds		
Linked File		Status	Reference Type		
i_struct_revised.rvt		Loaded	Overlay	

Note that the revised file has replaced the previous link.
This is similar to when a sub-contractor or other consultant emails you an updated file for use in a project.
Press **OK**.

25.

Coordination Review   Reconcile Hosting

Use Current Project

Select Link

Activate the **Collaborate** ribbon. Select **Coordination Review→Select Link** from the Coordinate panel. Select the linked file in the drawing window.

26.

Message	Action
New/Unresolved	
Levels	
Maintain Position	
Level moved by 0' - 6"	Postpone
i_struct_revised.rvt : Levels : Level : Roof TOS : id 185763	
Levels : Level : Roof : id 378728	
Level moved by 0' - 6"	Postpone
i_struct_revised.rvt : Levels : Level : Level 2 TOS : id 32567	
Levels : Level : Level 2 : id 9946	

A dialog appears.
Expand the notations so you can see what was changed.

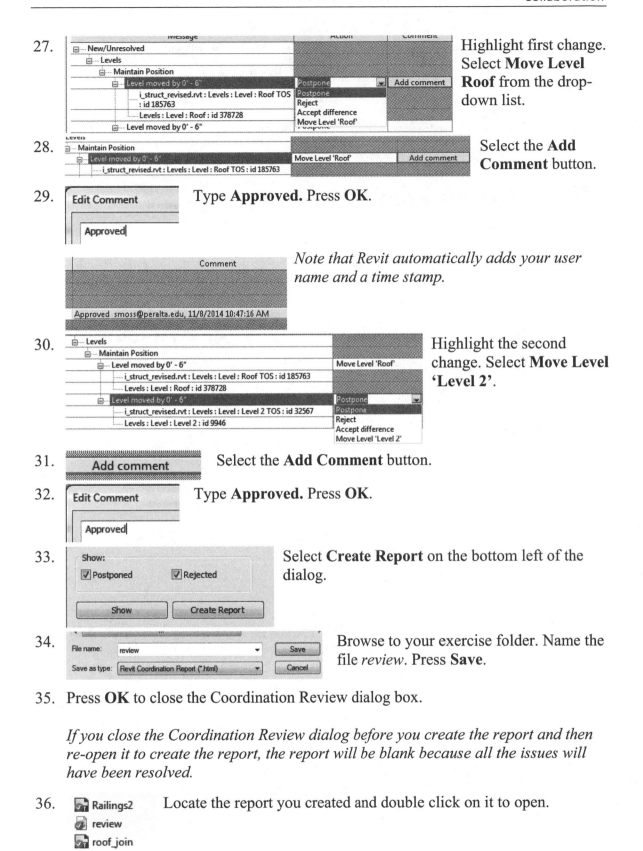

27. Highlight first change. Select **Move Level Roof** from the drop-down list.

28. Select the **Add Comment** button.

29. Type **Approved.** Press **OK**.

Note that Revit automatically adds your user name and a time stamp.

30. Highlight the second change. Select **Move Level 'Level 2'**.

31. Select the **Add Comment** button.

32. Type **Approved.** Press **OK**.

33. Select **Create Report** on the bottom left of the dialog.

34. Browse to your exercise folder. Name the file *review*. Press **Save**.

35. Press **OK** to close the Coordination Review dialog box.

If you close the Coordination Review dialog before you create the report and then re-open it to create the report, the report will be blank because all the issues will have been resolved.

36. Locate the report you created and double click on it to open.

37. This report can be emailed or used as part of the submittal process.

Revit Coordination Report

In host project

New/Unresolved	Levels	Maintain Position	Level moved by 0' - 6"	Levels : Level : Level 2 : id 9946 i_struct_revised.rvt : Levels : Level : Level 2 TOS : id 32567	Approved smoss@peralta.edu, 11/8/2014 10:48:09 AM
New/Unresolved	Levels	Maintain Position	Level moved by 0' - 6"	Levels : Level : Roof : id 378728 i_struct_revised.rvt : Levels : Level : Roof TOS : id 185763	Approved smoss@peralta.edu, 11/8/2014 10:47:16 AM

38. Close the file without saving.

Exercise 10-2
Interference Checking

Drawing Name: **c_interference_checking.rvt**
Estimated Time to Completion: 30 Minutes

Scope
Describe Interference Checks
Check and fix interference conditions in a building model
Generate an interference report

Solution

1. [image of Interference Check tool]   Activate the **Collaborate** ribbon.
   Select the **Interference Check→ Run Interference Check** tool on the Coordinate panel.

2. 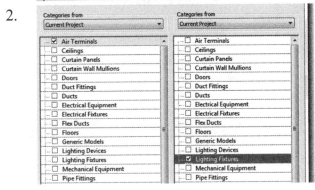   Enable **Air Terminals** in the left panel.
   Enable **Lighting Fixtures** in the right panel.

If you select all in both panels, the check can take a substantial amount of time depending on the project and the results you get may not be very meaningful.

3. Press **OK**.

4. Highlight the Lighting Fixture in the first error.

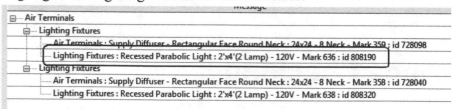

5. 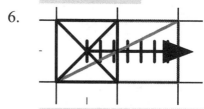 Press the **Show** button.

6. 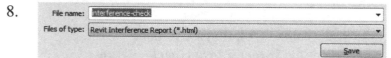 The display will update to show the interference between the two elements.

7. 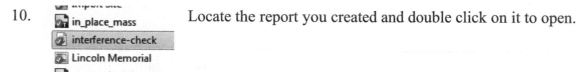 Select **Export**.

8.  Browse to your exercise folder.
   Press **Save**.

9. Press **Close** to close the dialog box.

10. Locate the report you created and double click on it to open.

11. This report can be emailed or used as part of the submittal process.

Interference Report

Interference Report Project File: C:\Users\emoss\SkyDrive\Revit Certification Guide 2014\Revit Certification exercises\ex10-2.rvt
Created: Tuesday, October 01, 2013 11:10:39 AM
Last Update:

A	B
1 Air Terminals : Supply Diffuser - Rectangular Face Round Neck : 24x24 - 8 Neck - Mark 358 : id 728040	Lighting Fixtures : Recessed Parabolic Light : 2'x4'(2 Lamp) - 120V - Mark 638 : id 808320
2 Air Terminals : Supply Diffuser - Rectangular Face Round Neck : 24x24 - 8 Neck - Mark 359 : id 728098	Lighting Fixtures : Recessed Parabolic Light : 2'x4'(2 Lamp) - 120V - Mark 636 : id 808190

End of Interference Report

12. Activate the **Collaborate** ribbon.
    Select the **Interference Check→ Show Last Report** tool on the Coordinate panel.

13. Highlight the first lighting fixture. Verify that you see which fixture is indicated in the graphics window. Close the dialog.

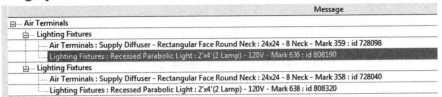

14. 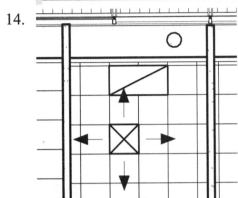 Move the lighting fixture to a position above the air terminal.
Arrange the lighting fixtures and the air diffuser so there should be no interference.

15. 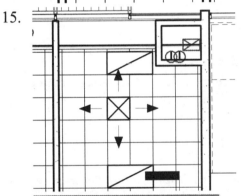 Move the second lighting fixture on the right so it is no longer on top of the air terminal.
Rearrange the fixtures in the room to eliminate any interference.

16. 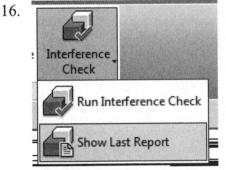 Activate the **Collaborate** ribbon.
Select the **Interference Check→ Show Last Report** tool on the Coordinate panel.

17. Refresh   Select **Refresh**.

18. The message list is now empty. Close the dialog box.

19. 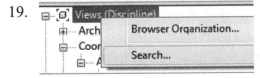 Highlight **Views** in the Project Browser.
Right click and select **Search**.

20.

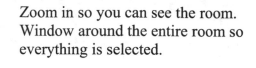

Type **Room 214**.
Press **Next**.

21.

Mechanical
  FP
    3D Views
      Room 214 3D Fire Protection
    HVAC

Press **Close** when the view is located.
Activate the **Room 214 3D Fire Protection view** located under Mechanical.

22.

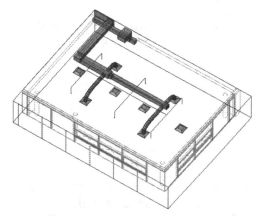

Zoom in so you can see the room.
Window around the entire room so everything is selected.

23.

e Interference Check

Run Interference Check

Activate the **Collaborate** ribbon.
Select the **Interference Check→ Run Interference Check** tool on the Coordinate panel.

24.

Categories from
Current Selection

☑ Air Terminals
☑ Duct Fittings
☑ Ducts
☑ Flex Ducts
☑ Mechanical Equipment
☑ Pipe Fittings
☑ Pipes
☑ Sprinklers

Categories from
Current Selection

☑ Air Terminals
☑ Duct Fittings
☑ Ducts
☑ Flex Ducts
☑ Mechanical Equipment
☑ Pipe Fittings
☑ Pipes
☑ Sprinklers

Note that the interference check is being run on the current selection only and not on the entire project.

Press **OK**.

25.

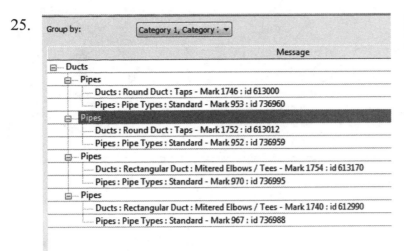

Review the report.
Press **Close**.

26. Close all files without saving.

In the certification exam, you might be asked to run an interference check and identify the id number of the element affected by interference. Can you locate the id number for the element with Mark Number 970 in the report?

Exercise 10-3
Using Shared Coordinates

Drawing Name: **Import Site.Dwg**
Estimated Time to Completion: 40 Minutes

Scope
Link an AutoCAD file
Set Shared Coordinates
Set Project North

Every project has a project base point ⊗ and a survey point △, although they might not be visible in all views, because of visibility settings and view clippings. They cannot be deleted.

The project base point defines the origin (0,0,0) of the project coordinate system. It also can be used to position the building on the site and for locating the design elements of a building during construction. Spot coordinates and spot elevations that reference the project coordinate system are displayed relative to this point.

The survey point represents a known point in the physical world, such as a geodetic survey marker. The survey point is used to correctly orient the building geometry in another coordinate system, such as the coordinate system used in a civil engineering application.

Solution

1. Start a new project file using the Architectural template.

2. Activate the **Site** floor plan.

3. 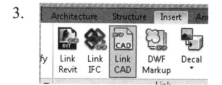 Activate the **Insert** ribbon.
    Select the **Link CAD** tool on the Link panel.

4. Select the *Import Site* drawing. Set Colors to **Preserve**. Set Import Units to **Auto-Detect**. Set Positioning to **Auto - Center to Center**. Press **Open**.

File name: Import Site
Files of type: DWG Files (*.dwg)

Colors: Preserve
Layers/Levels: All
Import units: Auto-Detect 1.000000
☑ Correct lines that are slightly off axis

Positioning: Auto - Center to Center
Place at: Level 1
☑ Orient to View
Open   Cancel

5.
Name   Import Site.dwg
Other
Shared Site   <Not Shared>

Select the imported site plan. In the Properties pane: Select the Shared Site button.

6.
◉ Acquire the shared coordinate system from "Import Site.dwg." This will modify the current model and all Named Positions of other linked models.

Record selected instance as being at Position:
Import Site.dwg : DefaultLocation   Change...

Enable **Acquire the shared coordinate system...**
Select **Change**.

7.
Define Location by:
Internet Mapping Service
Project Address:
800 Fallon Street, Oakland, CA 94607, USA   Search
Weather Stations:
4738 (5.60 miles away)
59482 (7.90 miles away)
4519 (9.70 miles away)
4737 (9.70 miles away)
4518 (13.70 miles away)
☐ Use Daylight Savings time
OK   Cancel   Help

Activate the Location tab. Enter the Project Address. Press **Search**, and then, press **OK**.

8. Reconcile   Press **Reconcile**.

9.

Note that the survey point and project base point shift position.

10.

Select the Project Base point.

This is the symbol that is a circle with an X inside.

Project Base Point
Shared Site:
N/S -267' 4 23/32"
E/W -171' 3 115/256"
Elev 126' 0"
Angle to True North 0.000°

11.

Project Base Point (1)	
Identity Data	
N/S	-267' 4 23/32"
E/W	-171' 3 115/256"
Elev	126' 0"
Angle to True North	90.000°

Change the Angle to Truth North to **90.00**.
Press **Apply**.
The site plan will rotate 90 degrees.

12.

Move the import site plan so that the corner of the rectangle is coincident with the project base point.
Use the **MOVE** tool.
Select the corner of the building and then the Project Base Point.

13.

Autodesk Revit 2017

Warning - can be ignored

Shared Sites in the link "Import Site.dwg" have been modified, but not saved
back to the link. Upon reopening, instances of the link will return to their last
Saved Positions. You can Save the link later via the Manage Links dialog.

| Show | More Info | Expand >> |

| Save Now | | OK | Cancel |

Select **OK**.

*If you select Save Now, this will modify
the linked file.*

14.

Use the PIN tool to prevent the site drawing from
shifting around.

15.

| Architecture | Structure | Insert | An |

Link Revit Link IFC Link CAD DWF Markup Decal

Link

Activate the Insert ribbon.
Select **Link Revit** from the Link panel.

16.

File name: i_shared_coords

Files of type: RVT Files (*.rvt)

Positioning: Manual - Base point

Select the *i_shared_coords* file.
Set the Positioning to: **Manual - Base point**.
Press **Open**.

17.

Place the Revit file to the left of the site
plan.

18.

Select the **Align** tool on the Modify panel on the Modify ribbon.

19.

Select the vertical line above the project base point.
Then select Grid line 1 on the shared_coords imported file.

20.

Change the display to wireframe, if necessary.
You can also move the linked Revit file up to locate the next alignment point.
Select the horizontal line on the project base point.
Then select Grid line E.
Cancel out of the ALIGN command.

21.

Select the *i_shared_coords* Linked Revit Model.

In the Properties pane:
Click on the Shared Site button.

22.

Enable **Record current position as
i_shared_coords.rvt**.
Press **Change**.

23. 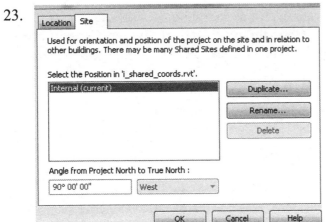 Press **OK**.
Press **OK** to close the Select Site dialog

24. Elevations (Building Elevation) Activate the North Elevation.
— East
— **North**
— South
— West

25. 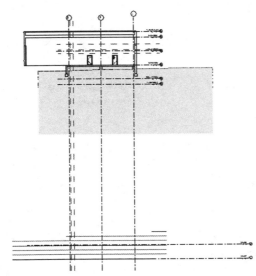 Level 1 and Level 2 are in the host project.
The other levels shown are in the Linked Revit model.

26. Views (all)
— Floor Plans
  — Level 1
  — Level 2
  — **Site**

Activate the **Site** plan.

27. 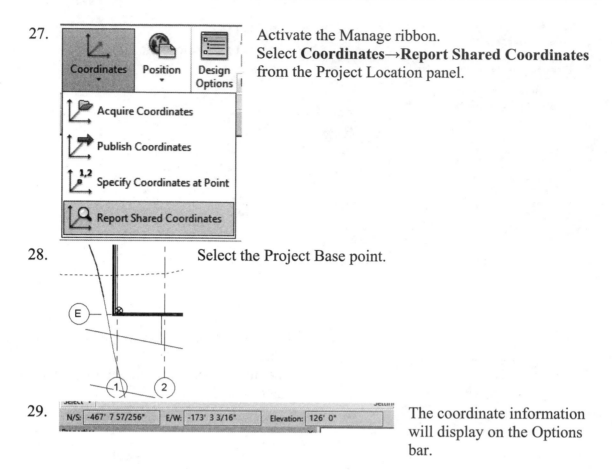 Activate the Manage ribbon.
Select **Coordinates→Report Shared Coordinates** from the Project Location panel.

28. Select the Project Base point.

29. | N/S: | -467' 7 57/256" | E/W: | -173' 3 3/16" | Elevation: | 126' 0" |

The coordinate information will display on the Options bar.

30. Close without saving.

Exercise 10-4
Worksets

Drawing Name: **i_Urban_house.rvt**
Estimated Time to Completion: 90 Minutes

Scope
Use of Worksets
Use of Revisions

Solution

1.

Before you can use Worksets, you need to set Revit to use your name.

Close any open projects.

Go to **Options**.

2.

Username

Elise

Select the **General** tab. Enter your first name in the Username text field. Press **OK**.

Username
smoss@peralta.edu
You are currently signed in to Autodesk 360. Your Autodesk ID is used as the username. If you need to change your username, you will need to sign out from Autodesk 360.
Sign Out of Autodesk 360

If you are using the Autodesk A360 account, you will see the user name assigned to that account. To use a different name, sign out from the A360 account.

3. Locate the *i_Urban_House* project. Press **Open**.

4. Collaborate   Select the **Collaborate** ribbon.

5. 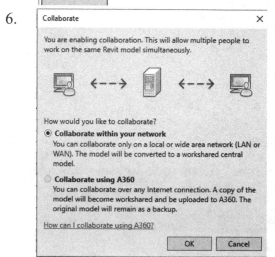 Select the **Collaborate** tool.
There will be a slight pause while Revit checks to see if you are connected to the Internet.

6. Enable **Collaborate within your network**. Press **OK**.

7. *Collaborate on A360 is only available for those users who purchase a separate subscription for Collaboration for Revit.*

8. Select **Worksets** on the Worksets panel.

9. 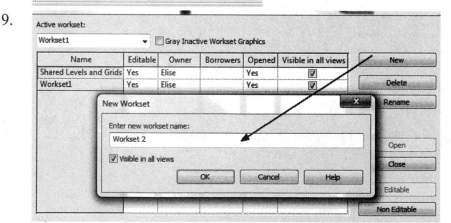 Create new worksets by pressing **New** and entering a name for the workset.

10.

Name	Editable	Owner	Borrowers	Opened
Shared Levels and Grids	Yes	Elise		Yes
Workset1	Yes	Elise		Yes
Workset 2	Yes	Elise		Yes

Identify the workset assigned to you.
Set Editable to **Yes**.
Your name should appear as the Owner of the workset.

11.

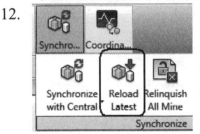

After each change, go to the Collaborate ribbon and select **Synchronize Now**. This ensures that your work is saved and that other team members can see your changes.

12.

If you select an element which has been modified, you may receive a prompt to load the latest version of the project. Select **Reload Latest** from the Collaborate ribbon to ensure you are working on the latest saved version of the project.

*You also want to **Reload Latest** after any breaks or long periods away from the project.*

13. Other team members may need to borrow elements assigned to your workset in order to make changes. You can Grant or Deny those requests.

Editing
Requests

Every 30 minutes or so, you should check your **Editing Requests**. Editing Requests also shows any pending requests you may have.

14.

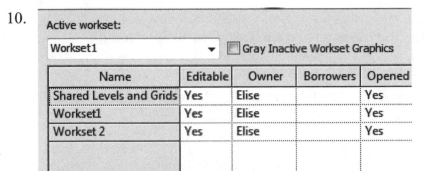

Highlight the view you are assigned to work on.
Right click and select **Duplicate View→ Duplicate**.

15.

Name: GROUND FLOOR - Workset 1

Rename the view with the [View Name]-Workset [Number].
Press **OK**.

16.
Revisions

Go to the View ribbon.
Select the **Revisions** tool on the Sheet Composition panel.

17. This is the table where revisions are listed as well as the team member assigned to make the change. *I have assigned changes to a Workset number. The instructor will assign each student a workset group to make changes.*

The Sequence Number is the same as the revision number.

Sequence	Numbering	Date	Description	Issued	Issued to	Issued by	Show
1	Numeric	08.02	Move Door 2 in the kitchen		Workset 1	Elise Moss	Cloud and Tag
2	Numeric	08.03	Change floor type to use gol	▦	Workset 1	Elise Moss	Cloud and Tag
3	Numeric	08.02	Demo Wall next to staircase	▦	Workset 1	Elise Moss	Cloud and Tag
4	Numeric	08.02	Change Opening to Staircas	▦	Workset 1	Elise Moss	Cloud and Tag
5	Numeric	08.03	Add door tags to all the doo	▦	Workset 2	Elise Moss	Cloud and Tag
6	Numeric	08.02	Change Door 17 Width to 4'-	▦	Workset 2	Elise Moss	Cloud and Tag
7	Numeric	08.02	Demo Door 19	▦	Workset 2	Elise Moss	Cloud and Tag
8	Numeric	08.02	Add a 36" x 84" Cased Openi	▦	Workset 3	Elise Moss	Cloud and Tag
9	Numeric	08.02	Demo the short horizontal w	▦	Workset 3	Elise Moss	Cloud and Tag
10	Numeric	08.03	Add a roof	▦	Workset 4	Elise Moss	Cloud and Tag
11	Numeric	08.03	Add a shade awning	▦	Workset 4	Elise Moss	Cloud and Tag
12	Numeric	08.03	Change Floor Type on Roof	▦	Workset 5	Elise Moss	Cloud and Tag
13	Numeric	08.03	Change Railing Type on Roo	▦	Workset 5	Elise Moss	Cloud and Tag

18.
Revision
Cloud

To place a revision cloud: Go to the Annotate ribbon.
Select **Revision Cloud** from the Detail panel.

Revisions should be placed on the existing view.

19.

Place a small cloud in the general area of the change.

20.

View "Floor Plan: GROUND FLOOR"

Elise

Seq. 4 - Change Opening to Staircase to 48" x 84"

4

08.02

Workset 1

Elise Moss

In the Properties pane:
In the Revision field, select the Sequence Number that is used for your change.
Press **OK**.

21. Select the **Green Check** on the Mode panel to **Finish Cloud**.

22. Activate the **Annotate** ribbon.
    Select **Tag by Category** from the Tag panel.

    Tag by
    Category

23. **No Tag Loaded for Object Type**
    You have not loaded a tag for this object type.
    Do you want to load a tag now?

    [Yes] [No]

    Pick the revision cloud.
    The first time you pick a revision cloud, you will be prompted to load the revision cloud tag. Press **Yes**.

24. Look in: Annotations
    Name
    North Arrow 2
    Revision Tag
    Room Tag

    Browse to the *Annotations* folder.
    Scroll down to locate the *Revision Tag* family.
    Press **Open**.

25.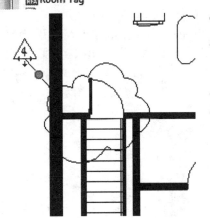

    The tag will display the sequence number assigned to the revision cloud.

    To change the sequence number, select the rev cloud, right click, select Element Properties, and change the Revision value.

26. Create a sheet for the view and name the sheet. The sheet should be named using [View] - [Workset Number].

27.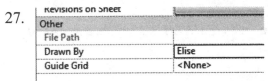

    Set Drawn By to your name.

28. Make the changes assigned to your workset.

Other Hints:

- Name any sheets you create. That way you can distinguish between your sheet and other sheets.
- Create a view that is specific to your changes or workset, so that you have an area where you can keep track of your work.
- Create one view for the existing phase and one view for the new construction phase. That way you can see what has changed. Name each view appropriately.
- Use View Properties to control the phase applied to each view.
- Check Editing Requests often.
- Duplicate any families you need to modify, rename, and redefine. If you modify an existing family, it may cause problems with someone else's workset.

Workset Checklist

1. Before you open any files, make sure your name is set in Options.
2. The main file is stored on the server. The file you are working on should be saved locally to your flash drive or your folder. It should be named Urban House [Your Name]. If you don't see your name on the file, you need to re-save or re-load the file.

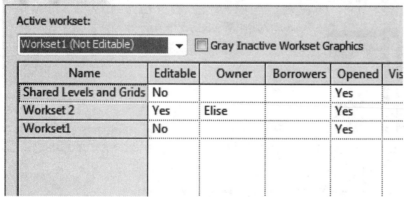

3. Go to Work sets and verify that your name is next to the Work set you are assigned.

4. Verify that the Active work set is the work set you are assigned.
5. Re-load the latest from Central so you can see all the updates done by other users.
5. When you are done working, save to Central – so other users can see your changes.
6. When you close, relinquish your work sets, so others can keep working.

The **Industry Foundation Classes (IFC)** data model is used to describe building and construction industry data.

It is a platform neutral, open file format specification that is not controlled by a single vendor or group of vendors. It is meant to promote data exchange between CAD vendors. It is a popular format in Europe and is promoted by several governments for use in their AEC projects. It is similar to the step/stp format used in mechanical design file exchanges. Revit has been improved for file import and export of IFC files.

Prior to importing or exporting with Revit, it is critical to set the IFC Options. The options allow you to map the different Revit elements appropriately.

To set the mapping options, use the Application Menu to go to Open→IFC Options.

Import IFC Options			

Default Template for IFC Import:

Import IFC Class Mapping: C:\Users\Elise\SkyDrive\Revit Certification Guide 2015\Certification Guide 2015 Exercises\StairimportIFCClassMapping

IFC Class Name	IFC Type	Revit Category	Revit Sub-Category
IfcAirTerminal		Air Terminals	
IfcAirTerminalType		Air Terminals	
IfcAnnotation		Generic Annotations	
IfcBeam		Structural Framing	
IfcBeamType		Structural Framing	
IfcBoiler		Mechanical Equipment	
IfcBoilerType		Mechanical Equipment	
IfcBuildingElementPart		Parts	
IfcBuildingElementPartType		Parts	
IfcBuildingElementProxy		Generic Models	

Then select which IFC Types to map to the appropriate Revit Category. This ensures the geometry comes in properly. You can also save and restore IFC mapping files.

Exercise 10-5

Import an IFC File

Drawing Name: Duplex_A_20110505
Estimated Time to Completion: 5 Minutes

Scope
Use of IFC Mapping files
Import IFC File

Solution

1.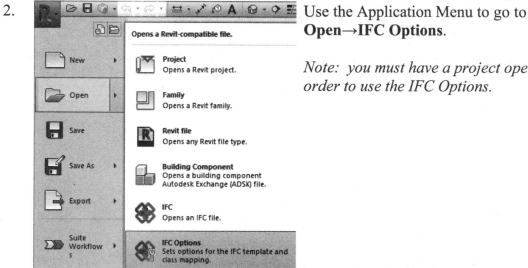

   Start a new project using the
   Architectural template.

2.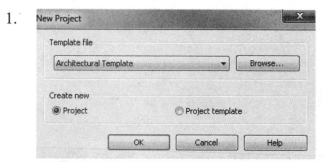

   Use the Application Menu to go to
   Open→IFC Options.

   *Note: you must have a project open in
   order to use the IFC Options.*

3.

   | Load... |
   | Standard |
   | Save As... |

   Select the **Load** button.

4. Select the *Duplex_A_20110505.ifc.sharedparameters* file. Press **Open**. Press **OK** to close the dialog.

Dds_BardNa.ifc.sharedparameters
Duplex_A_20110505.ifc.sharedparameters
Staircase.ifc.sharedparameters

5. 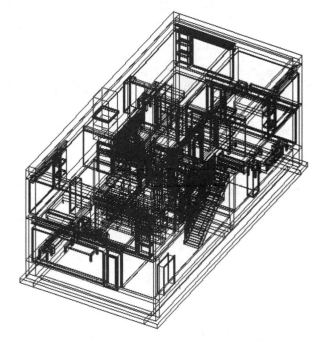 Select **Link IFC** on the Insert ribbon.

Link
IFC

6. 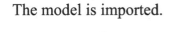 Select *Duplex_A_20110505.* Press **Open.**

File name: Duplex_A_20110505
Files of type: IFC Files (*.ifc)

7. The model is imported.

8. Close without saving.

Certified Professional Practice Exam

1. Worksharing allows team members to:

    A. Share families from different projects
    B. Share views from different projects
    C. Work on the same parts of a project simultaneously
    D. Work on different parts of the same project

2. Set Phase Filters for a view in the:

    A. Properties pane
    B. Design Options
    C. Ribbon
    D. Project Browser

3. A rectangular column family comes with three types: 18″ x 18″, 18″ x 24″, and 24″ x 24″. To add a 6″ x 6″ type, you:

    A. Duplicate, rename and modify the instance parameters
    B. Duplicate, rename and modify the type parameters
    C. Rename and modify the type parameters
    D. Duplicate, rename the family and add a type.

4. A door _____ a wall.

    A. hosts
    B. defines
    C. is hosted by
    D. is defined by

5. The Coordination Review tool is used when you use:

    A. Worksets
    B. Linked Files
    C. Interference Checking
    D. Phase

Answers
  1) D; 2) A; 3) D; 4) C; 5) B

About Us

SDC Publications specializes in creating exceptional books that are designed to seamlessly integrate into courses or help the self learner master new skills. Our commitment to meeting our customer's needs and keeping our books priced affordably are just some of the reasons our books are being used by nearly 1,200 colleges and universities across the United States and Canada.

SDC Publications is a family owned and operated company that has been creating quality books since 1985. All of our books are proudly printed in the United States.

Our technology books are updated for every new software release so you are always up to date with the newest technology. Many of our books come with video enhancements to aid students and instructor resources to aid instructors.

Take a look at all the books we have to offer you by visiting SDCpublications.com.

NEVER STOP LEARNING

Keep Going

Take the skills you learned in this book to the next level or learn something brand new. SDC Publications offers books covering a wide range of topics designed for users of all levels and experience. As you continue to improve your skills, SDC Publications will be there to provide you the tools you need to keep learning. Visit SDCpublications.com to see all our most current books.

Why SDC Publications?

- Regular and timely updates
- Priced affordably
- Designed for users of all levels
- Written by professionals and educators
- We offer a variety of learning approaches

TOPICS
3D Animation
BIM
CAD
CAM
Engineering
Engineering Graphics
FEA / CAE
Interior Design
Programming

SOFTWARE
Adams
ANSYS
AutoCAD
AutoCAD Architecture
AutoCAD Civil 3D
Autodesk 3ds Max
Autodesk Inventor
Autodesk Maya
Autodesk Revit
CATIA
Creo Parametric
Creo Simulate
Draftsight
LabVIEW
MATLAB
NX
OnShape
SketchUp
SOLIDWORKS
SOLIDWORKS Simulation